Y0-BTD-977

LASER ABSTRACTS

Volume 1

LASER ABSTRACTS

Volume 1

by
Prof. A.K. Kamal

Director, Quantum Electronics Laboratory
School of Electrical Engineering
Purdue University
Lafayette, Indiana

PLENUM PRESS
NEW YORK
1964

Author cross-index and subject index prepared
by the Plenum Press editorial staff

Library of Congress Catalog Card No. 64-20745

©1964 Plenum Press
A Division of Consultants Bureau Enterprises, Inc.
227 West 17th Street
New York, N.Y. 10011

Printed in the United States of America

To My Students

The field of lasers has grown at an extraordinary rate since the first successful experiments were performed by Maiman in 1960. The number of articles published in the field has also increased to such a point that many organized bibliographies have appeared during the past year.

This book is an attempt to present not only the bibliography but also abstracts, which should be of great help to the students, scientists, and engineers working in this field.

I am indebted to various of my colleagues for collecting material and for reading the manuscript. Special thanks must go to David Neal and K. B. Das for their sincere help. It is a pleasure to acknowledge the permission granted by the publishers of various journals and books for abstracting. Finally I wish to thank Plenum Press for very fine work in editing and printing.

A.K.K.

CONTENTS

1. Experiments on a Partially Shielded Ruby Laser Rod. R. L. Aagard, Proc. IRE, Vol. 50, pp. 2374-2375, November 1962.

   The effects of varying the length of the ruby which is pumped and the pumping energy per unit length are investigated.

2. Pumping Characteristics of a Partially Shielded Pulsed Ruby Laser. Roger L. Aagard, Honeywell Research Center, Hopkins, Minnesota, Lasers and Applications Symposium, Ohio State University, November 1962.

   A ruby laser rod has been optically pumped with a uniform flux per unit length while a portion of the rod was shielded from pumping radiation. The threshold for laser action has been measured as a function of the length of rod shielded at 300°, 200°, and 100°K. These results indicate the expected behavior at 300° and 200°K, but at 100°K the shielded portion of the rod appears to be pumped beyond saturation by the externally pumped portion of the rod.

3. Determination of the Diffraction Loss in a Pulsed Ruby Laser. R. L. Aagard, J. Opt. Soc. Am., Vol. 52, p. 1319, 1962.

   A method has been devised whereby diffraction loss in a ruby-laser crystal can be determined from a set of threshold measurements.

4. Emission Pattern of Ruby Laser Output. R. L. Aagard, D. L. Hardwick, and J. F. Ready, Appl. Optics, Vol. 1, pp. 537-538, July 1962.

   Composite rod optical masers exhibit a lowered threshold for laser action because of concentration of the pumping light along the axis of the rod by refractive effects. It is the purpose of this note to point out that a similar concentration of pumping light exists in solid ruby rods yielding an output of stimulated emission that occurs preferentially from the center portion of the rod.

5. Measurements of the Output from a Ruby Laser with a Central Hole in One of the End Mirrors. R. L. Aagard, J. Appl. Phys., Vol. 33, pp. 2842-2843, September 1962.

   The average transmission of one of the mirrors has been varied by changing the diameter of a hole in the center of one

1

of the aluminum end coatings. Measurements of the output of the laser as a function of the average transmission with fixed input energy show that the output is maximum for a particular value of transmission. The optimum transmission depends upon the input energy. A relationship between the optimum transmission and threshold for stimulated emission is shown.

6. Thermal Tuning of Ruby Optical Maser. I. D. Abella and H. Z. Cummins, J. Appl. Phys. (Correspondence), Vol. 32, pp. 1176-1177, June 1961.

   The oscillation wavelength in ruby maser is investigated as a function of temperature.

7. Effects of Temperature on Ruby Laser Mode Sequences. M. J. Adamson, T. P. Hughes, and K. M. Young. A.E.I. Research Laboratory, Aldermaston, Berkshire, U. K., Third International Symposium on Quantum Electronics, Paris, France, February 1963.

   Further time-resolved studies have been made of the spikes of output light from a ruby laser, using a rotating mirror camera and a Fabry-Perot etalon. The effects of temperature on the longitudinal and transverse mode sequences observed are described and interpreted.

8. Contributions to the Study of Lasers with Gases. I. Agirbiceanu, L. Blanaru, A. Agafitei, V. Vaciliu, I. Popescu, N. Ionescu and V. Velculescu, Institute of Atomic Physics, Bucharest, Rumania, Third International Symposium on Quantum Electronics, Paris, France, February 1963.

   Construction of an optimum laser device is discussed, based on multilayer mirrors, laser mountings, and procedures for optical adjustment. Theoretical aspects of energy transfer collisions in alkaline vapor and mixtures of gases are considered. It is shown that the variation of the pressure parameter may lead to an antipumping effect. Possibility of laser action in $Hg:N_2$ mixture is investigated.

9. Photodiode Detection. L. K. Anderson, Polytechnic Institute of Brooklyn Symposia Series, XIII, Optical Masers, April 1963.

   PIN junction photodiodes have been used as low-level detectors of light amplitude modulated at microwave frequencies.

2

Important factors in determining the detection efficiency in this application are effective quantum efficiency, electron and hole transit times, and diode impedance. The operation of a surface-junction germanium PIN diode will be discussed in terms of these factors and its performance compared with detectors of the photo-emissive and photo-conductive type.

10. Resonant Frequency Mixing Effects in Distributed Media. H. G. Anderson, H. Welling, and K. D. Moller, U. S. Army Electronics Research and Development Laboratory, Fort Monmouth, N. J. Third International Symposium on Quantum Electronics, Paris, France, February 1963.

> This paper describes frequency mixing effects in the milli-meter-wave region and theoretical expressions for the mixing efficiency are derived by using the density matrix formalism. In particular, the power dependence of the mixing efficiency is discussed for the two cases where the three frequencies w_1 (thermal relaxation), w_2 (relaxation) and $w_1 + w_2$ are either slightly off resonance or coincide exactly with the resonances of the particle system.

11. Use of a Michelson Interferometer to Determine Luminescence Spectra of Optical Maser Materials. E. Archbold and H. A. Gebbie, Proc. Phys. Soc., Vol. 80, pp. 793-794, 1962.

> The results of experimental tests for improved infrared spectra through the use of a Michelson interferometer with Fourier transformation are reported.

12. Thermal Effect and Power Enhancement in a He-Ne Optical Maser. F. T. Arecchi, Laboratori C.I.S.E., Milan, Italy, Third International Symposium on Quantum Electronics, Paris, France, February 1963.

> The main result of a study of the He-Ne laser is the evidence of an enhancement in the output power when increasing the temperature of the discharge. The effect can be exploited in order to increase the power range of the He-Ne laser in some applications.

13. Interactions between Light Waves in a Nonlinear Dielectric. J. A. Armstrong, N. Bloembergen, J. Ducuing, and P. S. Pershan, Phys. Rev., Vol. 127, pp. 1918-1939, September 1962.

The induced nonlinear electric dipole and higher moments in an atomic system, irradiated simultaneously by two or three light waves are calculated by quantum-mechanical perturbation theory. An important permutation symmetry relation for the nonlinear polarizability is derived and its frequency dependence is discussed. Nonlinear microscopic properties are related to an effective macroscopic nonlinear polarization. Explicit solutions are obtained for the coupled amplitude equations.

14. Influence of Phase Fluctuations and Spatial Variations in Nonlinear Optical Processes. J. A. Armstrong, N. Bloembergen, J. Ducuing, and P. S. Pershan, Bull. Am. Phys. Soc., II, Vol. 8, p. 233, March 1963.

> If laser beams used in harmonic generation and light mixing experiments contain many modes, fluctuations in the relative phase between these modes will produce variations in the harmonic intensity. A given mode at the second harmonic frequency may be generated by many pairs of fundamental modes. The theory of random processes in nonlinear devices has been applied to nonlinear optical processes.

15. Experimental Observations of Random Fluctuations in Optical Harmonic Generation. J. A. Armstrong and J. Ducuing, Bull. Am. Phys. Soc., II, Vol. 8, p. 233, March 1963.

> Data are presented showing that in harmonic generation using standard ruby lasers there is not a one-to-one correlation between the amplitude of laser spikes and the amplitudes of corresponding second harmonic spikes.

16. Splitting of the Ground State Levels of Ruby by an External Electric Field. J. D. Artman and J. C. Murphy, Bull. Am. Phys. Soc., II, Vol. 7, p. 14, January 1962.

> In ruby the Cr ion enters the Al_2O_3 lattice substitutionally in two nonequivalent sites which are related to each other by an inversion operation. These sites become nonequivalent upon application of an electric field. Using EPR, the splitting of the ground states has been observed.

17. Small-Angle Scatter in Solid State Optical Masers. J. G. Atwood, N. I. Adams, G. W. Dueker, and W. Gratz. J. Opt. Soc. Am., Vol. 52, p. 595, May 1962

The ability to predict the angular distribution of light emitted by a solid-state laser as a function of pumping power from measurements on the material of which it is made is investigated.

18. Light Source for Pumping of Continuous Solid-State Lasers. H. J. Auvermann, P. H. Keck, and C. E. White, Bull. Am. Phys. Soc., II, Vol. 7, p. 330, April 1962.

A light source is described which consists of an Osian XBO 1600-W high-pressure Xe arc lamp and an ellipsoidal mirror with a 24-in. diameter and an opening of f/0.4. The available flux density is sufficient to continuously pump any solid-state material which to date has been shown to display laser action in a flash unit.

19. Measuring Laser Output with Rat's Nest Calorimeter. R. M. Baker, Electronics, Vol. 36, pp. 36-38, February 1963.

Pulsed output energy of a laser beam is trapped and absorbed in a bundle of fine, insulated copper wire, whose change in resistance is proportional to the energy absorbed and is practically independent of the distribution of energy within the unit.

20. Mode Selection and Enhancement with a Ruby Laser. J. A. Baker and C. W. Peters, Appl. Optics, Vol. 1, p. 674, September 1962.

In this experiment the angular spread of the emission is limited by an aperture in an external optical system through which the standing wave passes. As a result the stimulated emission is limited to the same angular spread and the intensity of the laser beam is increased because of the suppression of the standing-wave pattern in off-axis directions.

21. Radio-Frequency Investigations of Gaseous Optical Masers. E. A. Ballik, Polytechnic Institute of Brooklyn Symposia Series, XIII, Optical Masers, April 1963.

Extension of basic techniques for observation of the output frequency spectrum of optical masers is described. The use of frequency and amplitude modulation techniques will allow the measurement of such parameters as active cavity Q and nonlinear effects in the medium. The spectral width of the output radiation is being investigated.

22. Optical and Cross Relaxation Maser Level Populations by Partial Distribution. W. A. Barker and J. D. Keating, Appl. Optics, Vol. 1, pp. 335-338, May 1962.

> The method of partial distributions is used to calculate the population distribution of three-level maser systems in which spontaneous emission and cross relaxation as well as thermal relaxation and induced absorption processes are included. A three-level optical maser is analyzed in which one spontaneous emission transition probability is included. The solution does not make the linear approximation to the Boltzmann factor.

23. On the Modulation of Optical Masers. F. S. Barnes, Proc. IRE (Correspondence), Vol. 50, pp. 1686-1687, July 1962.

> The effects of distorting the maser medium and the effective length of the resonant structure are discussed with regard to Stark and Zeeman modulation of optical masers.

24. Resonance of the Fabry-Perot Laser. S. R. Barone, AD293401, lv, October 1962.

> The optical mode structure of a Fabry-Perot interferometer-resonator composed of two infinite strip mirrors is investigated from the point of view of the general theory of nonspectral resonances. The classical description is supplemented by a discussion of the transverse resonance behavior.

25. Resonance of the Fabry-Perot Laser. S. R. Barone, J. Appl. Phys., Vol. 34, pp. 831-843, April 1963.

> The optical mode structure of a Fabry-Perot interferometer-resonator composed of two infinite strip mirrors is investigated from the point of view of the general theory of nonspectral resonances.

26. Population Inversion in a Discharge in a Mixture of Two Cases. N. G. Basov and O. N. Krokhin, Appl. Optics, Supplement 1, pp. 80-82, 1962.

> Conditions are considered for the formation of a population inversion in a gas discharge in a two-gas mixture, the atoms of which have equal energy levels.

27. Negative Absorption Coefficient at Indirect Transitions in Semiconductors. N. G. Basov, O. N. Krokhin, and J. M. Popov, pp. 496-506 in Advances in Quantum Electronics, J. R. Singer, ed., Columbia University Press, New York, 1961.

> A detailed study is carried out using the authors' method for obtaining the states with a negative temperature in semiconductors. The method is shown to require relatively small densities of excitation.

28. Theory of a Molecular Generator and Power Amplifier. N. G. Basov and A. M. Prokhorov, Doklady Akademii Nauk SSSR, Vol. 101, pp. 47-49, 1955.

> The analysis of the molecular generator operation is carried out by statistical quantum-mechanical methods with the aid of dispersion theory and by taking into account the saturation effect.

29. Optical Mixing. M. Bass, P. A. Franken, A. E. Hill, C. W. Peters, and G. Weinreich, Phys. Rev. Lett., Vol. 8, p. 18, January 1962.

> The observation of the sum frequency in the near-ultraviolet of two ruby laser beams of different frequencies coincident simultaneously upon a crystal of triglycine sulfate is reported.

30. Cathodoluminescent Optical Maser Pumping. C. W. Baugh, Jr., and J. W. Ogland, J. Opt. Soc. Am., Vol. 52, p. 602, May 1962.

> Studies and measurements have been made of cathodoluminescence for optical maser pumping. The topics considered are conversion efficiency, spectral matching, light flux control, saturation radiance, cooling and optical efficiency.

31. Quartz Ultraviolet Lasers. C. H. Becker, G. C. Cox, and D. B. McLennan, Proc. IEEE (Correspondence), Vol. 51, pp. 358-359. February 1963.

> In the course of quartz optical phono-maser research, stimulated emission of ultraviolet and violet optical photons in quartz have been observed.

32. Modulation Effects in Optical Pumping. F. D. Bedard, Laboratory

for Physical Sciences, College Park, Md., Third International Symposium on Quantum Electronics, Paris, France, February, 1963.

> Experiments have been performed in optical pumping with rubidium using large modulation indices with magnetic modulation. For modulation frequencies large compared with the resonance line width, the side band structure matches very well with that expected. However, for frequencies of value comparable with the line width no sidebands seem to appear and, in fact, the line only increases in size.

33. The Helium-Neon Laser as a Quantum Counter at 3.39 Microns. W. E. Bell, A. Bloom, and R. C. Rempel, Spectra-Physics, Mountain View, California, Third International Symposium on Quantum Electronics, Paris, France, February 1963.

> Inhibition of visible emission at 6328 A caused by the higher-gain infrared transition (3.39 microns) is used to turn the 6328 A laser into a sensitive detector of externally applied 3.39-micron radiation from another laser.

34. Radiative Lifetimes and Collision Transfer Cross Sections of Excited Atomic States. W. R. Bennett, Jr.,pp. 28-43 in Advances in Quantum Electronics, J. R. Singer, ed., Columbia University Press, New York, 1961.

> An experimental procedure for determining the actual decay rates of a system of levels is described. Data are taken as a function of pressure and used both to check approximate calculations of the Bates and Dangaart type and to determine the magnitude of the excitation transfer cross sections involved.

35. Hole Burning Effects in a He-Ne Optical Maser. W. R. Bennett, Jr., Phys. Rev., Vol. 126, pp. 580-593, April 1962.

> A study has been made of pulling effects by the amplifying media on the TEM_{00} modes of a helium-neon maser using a circular plane mirror Fabry-Perot cavity in which the mirror separation was known with precision. Approximate expressions are derived for mode pulling in homogeneously and inhomogeneously broadened optical masers. The case of Lorentzian holes burned in a Gaussian line is treated specifically.

36. Gaseous Optical Masers. W. R. Bennett, Jr., Appl. Optics, Supplement 1, pp. 24-62, 1962.

> A detailed review of the present knowledge of gaseous optical masers is given. Included are: a summary of basic general considerations, a review of the dominant excitation mechanisms which have been used to produce population inversions in gas lasers, consideration of various aspects of operating cw lasers, and spectral characteristics and mode pulling effects.

37. Dissociative Excitation Transfer and Optical Maser Oscillation in Ne · O_2 and Ar · O_2 rf Discharges. W. R. Bennett, Jr., W. L. Faust, R. A. McFarlane, and C. K. Patel, Phys. Rev. Lett., Vol. 8, pp. 470-473, June 1962.

> Excitation studies conducted in low-pressure rf discharges containing noble gases with varying amounts of oxygen impurity are summarized. The dominant mechanisms by which radiative excited states of atomic oxygen are produced under these conditions involve quasi-resonant transfer of energy from noble gas metastables to repulsive, neutral excited states of the oxygen molecule. Application of these results to Ne and Ar mixtures has permitted continuous optical maser operation.

38. Magnetostrictively Tuned Optical Maser. W. R. Bennett, Jr., and P. J. Kindlmann, Rev. Sci. Instr., Vol. 33, pp. 601-605, June 1962.

> A Fabry-Perot helium-neon maser has been constructed in which both angular adjustments and plate separation may be controlled through magnetostrictively produced distortion in a rigid frame. The maser has been designed with the eventual use of negative feedback frequency stabilization in mind. Orthogonality is maintained over a range of about 15". Mechanical design considerations and construction details are discussed.

39. Collision Cross Sections and Optical Maser Considerations for Helium. W. R. Bennett, Jr., and P. J. Kindlmann, Bull. Am. Phys. Soc., II, Vol. 8, p. 87, January 1963.

> Measurements of excited-state decay rates as a function of helium pressure are reported.

40. Possible Use of Semiconductors in Lasers. C. Benoit à la Guillaume and (Mme.) Tric, J. Phys. Rad. (France), Vol. 22, pp. 834-836, December 1961 (in French).

> Laser action may be possible by population inversion or by the use of second-order transitions with the simultaneous emission of a phonon and a photon.

41. Cross Sections for the De-Excitation of Helium Metastable by Impurities. E. E. Benton and W. W. Robertson, Bull. Am. Phys. Soc., II, Vol. 7, p. 114, February 1962.

> The effects of small amounts of argon, krypton, xenon, neon, nitrogen, and hydrogen on the destruction frequencies of the 21s and 23s metastable states of helium are given.

42. Stationary Modes in Optic and Quasioptic Cavities. L. Bergstein and H. Schacter, Polytechnic Institute of Brooklyn Symposia Series, XIII, Optical Masers, April 1963.

> The resonant modes of optic and quasioptic cavities are found by means of an infinite orthogonal series expansion. The terms of the series can be interpreted as Fraunhofer diffraction patterns of apertures having the same geometry as the end reflectors of the cavity, each of the patterns being centered on the particular Fresnel zone. The eigenvalues are found as a function of the Fresnel number. Approximate expressions for the eigenvalues are found for the case of plane mirrors of rectangular and circular shape.

43. A Total-Reflection Solid-State Optical Maser Resonator. L. Bergstein and C. Shulman, Proc. IRE, Vol. 50, p. 1833, August 1962.

> A novel resonator and output coupling configuration for an optical maser is described. This configuration differs from others in current use in that it uses total internal reflection to reflect the light beam and frustrated total reflection for output coupling.

44. Coherence Studies of Emission from a Pulsed Ruby Laser. D. A. Berkley and G. J. Wolga, Cornell University, Ithaca, N. Y., Third International Symposium on Quantum Electronics, Paris, France, February 1963.

> A study of the cross-correlation function of far-field ruby

laser emission is reported. Clear interference fringes are obtained from two slits that are excited with a variable time delay.

45. Coherence Studies of Emission from a Pulsed Ruby Laser. D. A. Berkley and G. J. Wolga, Phys. Rev. Lett., Vol. 9, pp. 479-482, December 1962.

    The study of the cross-correlation function of the far-field ruby laser emission is reported. Clear interference fringes were obtained from two slits that were separately illuminated by different parts of the laser beam cross section and with a variable delay time between the illuminating of one slit and the subsequent illuminating of the other slit by a portion of the same wave front.

46. Possibility of Lasers Using Semiconductors. M. Bernard and G. Duraffourg, J. Phys. Rad. (France), Vol. 22, pp. 836-837, December 1961 (in French).

    Thermodynamic considerations are applied to systems in which emission of two bosons is possible, one stimulated and the other thermalized. An expression for the difference between the yield of coherent radiation and the larger Carnot yield is derived. Practical possibilities in semiconductors are discussed.

47. Possible Semiconductor Lasers. M. Bernard and G. Duraffourg, Department Physique Chimie Metallurgie C. N. E. T., Issy-les-Moulineaux, Seine, France, Third International Symposium on Quantum Electronics, Paris, France, February 1963.

    Laser effect in semiconducting materials is considered for different radiative mechanisms: band to band, band to impurity level, annihilation of free excitons, and annihilation of trapped excitons. Impurities in germanium and silicon have been particularly studied. The oscillation criterion is calculated for several models and first experimental results are reported.

48. Quantum Electrodynamic Prediction of the Envelope Modulation of Maser Beams. P. M. Bevensee, Proc. IEEE (Correspondence), Vol. 51, pp. 215-216, January 1963.

    The equations for the field and molecular energies are

developed as a function of time and predict a damping and modulation frequency as given by Singer and Wang.

49. Requirements of a Coherent Laser Pulse Doppler Radar. G. Biernson and R. F. Lucy, Proc. IEEE, Vol. 51, pp. 202-213, January 1963.

> The use of coherent detection can theoretically allow optical radar systems employing laser transmitters to achieve considerably improved receiver sensitivity, particularly in conditions of high background radiation. Many practical factors limit sensitivity. An efficient coherent optical radar would likely require a pulse width of less than 10 microseconds and a spectral line width of less than 10 mc.

50. Rectangular Optical Dielectric Waveguides as Lasers. V. R. Bird, D. R. Carpenter, P. S. McDermott, and R. L. Powell, IBM Federal Systems Division, Oswego, New York, Lasers and Applications Symposium, Ohio State University, November 1962.

> Techniques have been developed with the preparation of the crystalline dielectric optical waveguides of known laser materials to cross-sectional dimensions as small as 50 by 25 microns and greater than 2.5 cm in length.

51. Optical Maser Studies. M. Birnbaum, AD293197, 23 pp., November 1962.

> Effort was expended in a fundamental study of useful or potentially useful materials in optical maser technology. Small crystals of beta-Ga_2O_3 and powders of alpha- and beta-Ga_2O_3 have been prepared and preliminary observations of their fluorescence have been made.

52. Pulsed Oscillations in Ruby Lasers. M. Birnbaum, T. Stocker, and S. J. Welles, Proc. IEEE, Vol. 51, pp. 854-855, 1963.

> The Statz-de Mars equations are revised into a form more appropriate for the three-level system, solutions of which are found to be in better agreement with the experimental results for ruby lasers.

53. Recombination Radiation in GaAs. J. Black, H. Lockwood, and S. Mayburg, J. Appl. Phys., Vol. 34, pp. 178-180, January 1963.

12

Recombination radiation in GaAs diodes has been found to
be proportional to the forward diode current at current levels
where the recombination is dominant, implying an efficient
conversion from injected electrons to band-edge photons. A
visible red radiation has been observed from these diodes at
high forward diode currents.

54. Les Laser. A. Blandin, L'Onde Electrique, Vol. 41, pp. 931-939,
November 1961.

Various aspects of recent developments in laser technology
are discussed (including the calculation of the Q of the Fabry-
Perot cavity, spatial coherence, types of lasers, and applications). A large portion of the material is based on work
done in the United States and reported in English language
articles.

55. Identification of Lasing Energy Levels by Spectroscopic Techniques.
E. J. Blau, B. F. Hockheimer, J. T. Massey, and A. G. Schulz,
J. Appl. Phys., Vol. 34, p. 703, March 1963.

The technique is based on the observation that some of the
processes by which energy levels are populated are independent of the lasing field.

56. Harmonic Light Waves. N. Bloembergen, Bull. Am. Phys. Soc.,
II, Vol. 7, p. 196, March 1962.

The existence of intense coherent light sources has made
possible the creation of light harmonics due to the inherent
nonlinearities in a dielectric medium. For the case in which
the phase velocities of the fundamental and only one harmonic
wave are nearly completely matched, analytical large-sig-
nal solutions for the power transfer between the two monochromatic plane waves has been obtained.

57. Some Theoretical Problems in Quantum Electronics. N. Bloembergen, Polytechnic Institute of Brooklyn Symposia Series, XIII,
Optical Masers, April 1963.

The adaptation of quantum-mechanical concepts from the
realm of radio- and microwaves to the optical region is
considered.

58. Observation of New Visible Gas Laser Transitions by Removal of

Dominance. A. L. Bloom, Appl. Phys. Lett., Vol. 2, pp. 101-102, March 1963.

> Dominance is eliminated by incorporating a prism into the laser resonator cavity.

59. Laser Operation at 3.39 Microns in a Helium-Neon Mixture. A. L. Bloom, W. E. Bell, and R. C. Rempel, Appl. Optics, Vol. 2, pp. 317-318, March 1963.

> The transition is shown to be from $^3S_2$ to $^3P_4$ in neon. Lasers have also been observed oscillating simultaneously on the visible and 3.39 micron lines.

60. Reception of Single-Sideband Suppressed-Carrier Signals by Optical Mixing. L. R. Bloom and C. F. Buhrer, Proc. IEEE, Vol. 51, pp. 610-611, April 1963.

> The authors describe a receiving apparatus in which sideband signals can be demodulated by optically mixing them with a reinserted carrier in a photodetector.

61. Design of a Microwave Frequency Light Modulator. R. H. Blumenthal, Proc. IRE, Vol. 50, pp. 452-456, April 1962.

> A high-Q resonator employing a uniaxial crystal of KDP as its dielectric medium is coupled to a microwave frequency generator tuned to the fundamental cavity mode. The method of design is fully described.

62. Fast Response Solid State PME Detector for Laser Signals. A. Boatright and H. Mette, AD286656, 16 pp., June 1962.

> A germanium device with response time below 0.1 microsecond is described.

63. Making Crystal Elements for Optical Masers. W. L. Bond, Rev. Sci. Instr., Vol. 33, pp. 372-375, March 1962; Appl. Optics, Supplement 1, pp. 114-117, 1962.

> Methods are given by means of which crystal maser rods have been made with the ends parallel to better than 10" and flat to better than one-tenth wavelength of visible light.

64. Observation of the Dielectric-Waveguide Mode of Light Propagation

in p-n Junctions. W. L. Bond, B. G. Cohen, R. C. Leite, and A. Yariv, Appl. Phys. Lett., Vol. 2, pp. 57-59, February 1963.

The observation of the dielectric-waveguide mode is based on the distribution of light energy across the face of the diode junction.

65. Pulsed Gaseous Maser. H. A. Boot and D. M. Clunie, Nature, Vol. 197, pp. 173-174, January 1963.

Optical maser oscillation has been obtained during the recombination period following a pulsed radio-frequency or dc discharge in helium mixed with either neon or carbon monoxide at wavelengths of 1.153 and 1.069 microns, respectively.

66. Ultrarapid Photography of the Emitting Surface of a Ruby Crystal Laser. J. C. Borie, M. Durand, and A. Orszag, Compt. Rend., Vol. 253, pp. 2215-2217, 1961 (in French).

Analysis of oscillograms of light from a ruby laser shows that luminous impulses of 0.1 microsecond duration recur at intervals of 0.5 microsecond. Photographs of the surface of the crystal show an irregular granular distribution of the light in spots of about 0.01 cm. The two images obtained by passing the light through a birefringent crystal show differences in intensity in the spots indicating local differences in polarization by the crystal.

67. Infrared Oscillations from $CaF_2 : U^{3+}$ and $BaF_2 : U^{3+}$ Masers. H. A. Bostik and J. R. O'Connor, Proc. IRE, Vol. 50, pp. 219-220, February 1962.

An analysis of the oscillations suggests that the initial pulse train in the output is either one mode or a small number of coupled modes, and that a second mode becomes resonant as the cavity temperature is raised.

68. A High-Energy Laser Using a Multielliptical Cavity. C. Bowness, D. Missio, and T. Rogala, Proc. IRE, Vol. 50, pp. 1704-1705, July 1962.

A four-ellipse cavity is described. It is found that the efficiency of coupling light into the cavity decreases as the number of cavities increases.

15

69. "Author's Comment." C. Bowness, D. Missio, and T. Rogala, Proc. IEEE, Vol. 51, p. 255, 1963.

   Author's comments on the use of multicavity lasers are described.

70. The Confocal Resonator for Millimeter Through Optical Wavelength Masers. G.D. Boyd, pp. 318-327 in Advances in Quantum Electronics, J. R. Singer, ed., Columbia University Press, New York, 1961.

   A resonator formed by two spherical reflectors separated by their common radius of curvature is suggested as an alternative to the plane-parallel Fabry-Perot interferometer.

71. Excitation, Relaxation, and Continuous Maser Action in the 2.613 Micron Transition of $CaF_2:U^{3+}$. G. D. Boyd, R. J. Collins, S. P. Porto, A. Yariv, and W. A. Hargreaves, Phys. Rev. Lett., Vol. 8, pp. 269-272, April 1962.

   An investigation of the 2.613-micron fluorescence of trivalent uranium in CaF_2 and the operation of a continuous solid-state maser using this transition are described.

72. Confocal Multimode Resonator for Millimetre through Optical Wavelength Masers. G. D. Boyd and J. P. Gordon, B.S.T.J., Vol. 40, pp. 489-508, March 1961.

   A plane-parallel Fabry-Perot interferometer could act as a suitable resonator. A resonator consisting of two identical concave spherical reflectors separated by any distance up to twice their common radius of curvature is considered.

73. Generalized Confocal Resonator Theory. G. D. Boyd and H. Kogelnik, B.S.T.J., Vol. 41, pp. 1347-1370, July 1962.

   The theory of the confocal resonator is extended to include the effect of unequal aperture size and unequal radii of curvature of the two reflectors. The latter is equivalent to a periodic sequence of lenses with unequal focal lengths. This treatment is in Cartesian coordinates, as used previously. In an appendix the modes and resonant formulas are written in cylindrical coordinates.

74. Defining the Coherence of a Signal. R. N. Bracewell, Proc. IRE,

Vol. 50, p. 214, 1962.

> From the discussion it follows that the degree of stationarity of a signal should enter into the framing of a definition of its coherence and that mention of the lapse time I should be made in statements about the coherence of a signal.

75. Optical Maser Detection by Microwave Absorption in Semiconductors. F. A. Brand, H. Jacobs, S. Weitz, and J. Strozyk, Proc. IEEE, Vol. 51, pp. 607-609, April 1963.

> This communication describes some results in a study of a ruby maser output using microwave semiconductor absorption techniques for optical detection.

76. Construction of a Gaseous Optical Maser Using Brewster Angle Windows. D. J. Brangaccio, Rev. Sci. Instr., Vol. 33, pp. 921-922, September 1962.

> This paper describes the construction techniques used to make a gaseous optical maser using windows oriented at the Brewster angle. The use of quartz for the gas discharge tube avoids gas contamination. The tube consists of two optical-quality glass windows held at the Brewster angle, mounted in hemispherical bulbs and connected together with a quartz tube through appropriate graded seals. A description is given of the materials, tools, and techniques used in construction.

77. Changes in the Absorption Spectrum of Substances under High-Intensity Coherent Illumination. J. Braunbeck, Varian A. G., Zurich, Switzerland, Third International Symposium on Quantum Electronics, Paris, France, February 1963.

> If a beam of coherent light is focused on a small area, electric field strengths of the order of 10^9 v/meter can be obtained. An apparatus has been built to study the influence of this strong electromagnetic field on the absorption spectrum in other wavelength regions.

78. Coherent Stimulated Emission from Organic Molecular Crystal. E. G. Brock, P. C. Savinszky, E. Hormats, H. C. Nedderman, D. Stirpe, and F. Unterleitner, J. Chem. Phys., Vol. 35, pp. 759-760, August 1961.

Conjugated organic molecules when incorporated in crystals may produce saturated maser materials. The use of such crystals in optical masers and optically controlled microwave masers is discussed.

79. Coupling between Samarium Ions and Colour Centers in the Calcium Fluoride: Samarium Laser System. P. F. Browne, Proc. Phys. Soc. A, Vol. 79, pp. 1085-1087, May 1962.

Factors affecting the laser performance of the crystals are determined by an analysis of the energy level diagram. It is proposed that the efficiency of the CaF_2:Sm system is due to the transfer of energy from color centers to Sm impurity ions. Thus, for efficient laser action crystals should be grown with excess Ca so as to include color centers.

80. Photon Beam Machining with Laser Generator. M.S. Bruma, C.N.R.S., Paris, Third International Symposium on Quantum Electronics, Paris, France, February 1963.

The laser principle extends and complements the electron beam method of machining, by exploiting the energy contained in a photon beam. An object which intercepts the beam, if it does not reflect it, will absorb the incident energy in the form of a thermal impulse localized on the impact surface. It is possible to forecast theoretically in specific cases the melting or vaporization conditions of an elementary volume of the object under consideration.

81. Xenon Flash Tubes for Laser Pumping. A. Buck, R. Erickson, and F. Barnes, University of Colorado, Boulder, Colorado, Third International Symposium on Quantum Electronics, Paris, France, February 1963.

A study of xenon flash tubes has been made to determine both the most efficient mode of operation for driving lasers and the design parameters required to obtain a high-intensity, high-efficiency optical source.

82. An Experimental Laser Ranging System. D. A. Buddenhagen, B. A. Lengyel, F. J. McClung, and G. F. Smith, IRE Int. Conv. Rec., Part 5, Vol. 9, pp. 285-290, 1961.

An experimental ranging system has been constructed using a ruby laser. A 0.4-milliradian beam width is obtained.

83. Frequency Shifts of Light Beams. C. Buhrer, L. Bloom, V. Fowler, D. Baird, and E. Conwell, General Telephone and Electronics Laboratories, Bayside, New York, Third International Symposium on Quantum Electronics, Paris, France, February 1963.

> Using the principle that a rotating birefringent plate, simulated by means of the Pockels electro-optic effect, will act upon a circularly polarized light beam to produce a separable component shifted in frequency, it is shown that this effect can be obtained with one crystal if light travels along a threefold symmetry axis of a crystal such as natural cubic zinc sulfide.

84. Single-Sideband Suppressed-Carrier Modulation of Coherent Light Beams. C. F. Buhrer, V. J. Fowler, and L. R. Bloom, Proc. IRE, Vol. 50, pp. 1827-1828, August 1962.

> A modulator employing the Pockels electro-optic effect in a pair of potassium dihydrogen phosphate crystals is reported.

85. Ruby Masers with Afocal Resonators. J. M. Burch, J. Opt. Soc. Am., Vol. 52, p. 602, May 1962.

> Maser operation has been achieved using a lens and a pair of plane reflectors. Because the system is afocal there is no walk-off effect and small wedge errors of the ruby are automatically compensated.

86. Optical Faraday Rotation and Microwave Interactions in Paramagnetic Salts. J. Q. Burgess and W. S. Chang, J. Opt. Soc. Am., Vol. 51, p. 477, April 1961.

> The effect of paramagnetic absorption in optical Faraday rotation of light can be used as a microwave and millimeter detector or as an optical polarization modulator for coherent radiation in maser and laser applications.

87. Directionality Effects of GaAs Light-Emitting Diodes: Part 1. G. Burns, R. A. Laff, S. E. Blum, F. H. Dill, Jr., and M. I. Nathan, IBM J. Res. Div., Vol. 7, p. 62, January 1963.

> The fabrication and operation of the diodes so that strong directional effects are obtained are discussed.

19

88. Correction to Line Shape in GaAs Injection Lasers. G. Burns and M. I. Nathan, Proc. IEEE, Vol. 51, p. 860, 1963.

> In the second sentence of the first paragraph on page 472, Proc. IEEE, Vol. 51, pp. 471-472, March 1963, the words "as well as highly directional" should be deleted.

89. Line Shape at 77°K of GaAs Injection Laser. G. Burns, M. I. Nathan, B.A. Jenkins, and G. O. Pettit, Bull. Am. Phys. Soc., Vol. 8, p. 88, 1963.

> Pulsed measurements were taken on GaAs diodes. At low currents one observes a line approximately 140 A wide at 8400 A. As the current is raised a sharp spike comes out of this wide line. The sharp spike comes out of the broad base at a fairly well-defined position, relatively independent of the current. It tends to come on the long-wavelength side of the broad base. High resolution has shown lines as narrow as 0.2 A.

90. Room-Temperature Stimulated Emission. G. Burns and M. I. Nathan, IBM J. Res. Dev., Vol. 7, pp. 72-73, January 1963.

> Line narrowing and directionality at room temperature in forward-biased GaAs p-n junctions are reported.

91. Quantum Efficiency of Ruby. G. Burns and M. I. Nathan, J. Appl. Phys., Vol. 34, pp. 703-704, March 1963.

> It is shown that to 240°C the quantum efficiency is independent of temperature. The fraction of the radiation that is emitted in the R lines is also measured.

92. Line Shape in GaAs Injection Lasers. G. Burns and M. I. Nathan, Proc. IEEE, Vol. 51, pp. 471-472, March 1963.

> Experiments to determine how the stimulated emission arises from wide lines, its effect on the wide lines and temperature dependence are described. The data for the first time confirm the time-independent solutions of the simple laser rate equations.

93. Investigations in Experimental and Theoretical Physics. F. A. Butayeva and V. A. Fabrikant, Izd. Akad. Nauk SSSR, Moscow.

A collection of papers in memory of G. S. Landsberg, Academy of Sciences of the USSR, Moscow.

94. Observation of Increased Power Output from Ne-He Optical Maser by Means of Externally Applied High-Voltage Pulsing. E. H. Byerly, J. Goldsmith, and W. H. McMahan, Proc. IEEE (Correspondence), Vol. 51, p. 360, February 1963.

   While conducting gas laser experiments using a pulsed rf exciter, it was observed that the output under pulsed conditions exceeded the cw power output by about 25%. Pulsed-mode operation minimizes the population of the Ne (1s) meta-stable state and the elevation of the gas temperature, phenomena inherent in continuous operation which limit power output.

95. An Optical Calorimeter for Laser Energy Measurements. J. A. Calviello, Proc. IEEE, Vol. 51, pp. 611-612, April 1963.

   A method for providing improved calibration for the carbon cone calorimeter is described.

96. Masers, Lasers, and the Ether Drift. C. W. Carnahan, Proc. IRE, Vol. 49, pp. 1576-1577, October 1961.

   The author suggests that an ether drift experiment of Cedar-holm et al. (Phys. Rev. Lett., Vol. 1, p. 342) gives a null result because it is not affected by an ether drift. From a consideration of the error in this experiment a new ether drift test employing an optical maser is proposed.

97. A Proposed First-Order Relativity Test Using Lasers. C. W. Carnahan. Proc. IRE, Vol. 50, p. 1976, September 1962.

   In the proposed ether drift test the effect to be measured is proportional to the first power of v/c, where v is the expected ether drift.

98. Modulation of Laser Output by Multiple Reflection Kerr-Effect on Thin Magnetic Films. D. Chen, Polytechnic Institute of Brooklyn Symposia Series, XIII, Optical Masers, April 1963.

   Numerical calculations using Argyre's analysis applied to the case of a ruby laser beam reflected once from a Ni-Fe film predict a very small modulation index. By arranging two such films face to face the laser beam may be multiply reflected, resulting in a considerable increase in the Kerr component. The effect of microwave modulation may be achieved by applying a microwave field at the ferromagnetic resonance frequency of the magnetic films which form the reflecting surfaces in a slow wave structure.

99. Production of a Single Pulse of Far-Infrared Radiation (160 Microns) Using Beats from a Single Ruby Laser. R. H. Christie, AWRE, Aldermaston, Berks, U. K., Third International Symposium on Quantum Electronics, Paris, France, February 1963.

An experimental system is described in which an intense single pulse of radiation is emitted from a ruby laser, using a heavily doped (0.55%) sample. Triggering of this output pulse is achieved by operating a Faraday cell within the optical cavity. Details of pulse shape and energy are given for the two component wavelengths of the laser emission. Conditions required for simultaneous emission at these two wavelengths are derived.

100. Optical Pumping of Lasers Using Exploding Wires. Charles H. Church, R. D. Haun, Jr., T. W. O'Keeffe, and T. A. Osial, Westinghouse Research Laboratories, Pittsburgh, Pennsylvania, Lasers and Applications Symposium, Ohio State University, November 1962.

The use of exploding wires for the optical pumping of lasers was investigated. On the basis of the data presently available, the effect on the radiant energy flux of changes in wire size, shape, and composition was determined. Saturation in the light output for increasing voltage was observed, holding other parameters constant. Various schemes for protecting the laser rod from the shock wave were tried. The characteristics of a ruby laser pumped with an exploded wire has been presented, along with some of the experimental techniques useful in these studies.

101. Optical Pumping of Lasers Using Exploding Wires. C. H. Church, R. D. Haun, Jr., T. A. Osial, and E. V. Sommers, Appl. Optics, Vol. 2, pp. 451-452, April 1962.

The results of the experiment indicate that the exploding

22

wire allows much higher pumping rates than flash lamps. As a source of light in the visible it appears to be less efficient. Its efficiency for pumping lasers could be enhanced appreciably by a better choice of laser conditions together with improved coupling to the laser rod of the light emitted by the plasma.

102. Optical Pumping of Lasers Using Exploding Wires. C. H. Church, R. D. Haun, Jr., T. A. Osial, and E. V. Sommers, J. Opt. Soc. Am., Vol. 52, p. 603, May 1962.

> The use of exploding wires as a pump source has the advantages of fast rise times and high peak power levels. A description is included of the equipment and the precautions necessary to avoid damage to the ruby by the shock wave from the exploding wire at high power levels.

103. Coaxial Laser Pumps. C. H. Church, Derek Ryan, and J. P. Lesnick, J. Opt. Soc. Am., Vol. 53, p. 514, 1963.

> A study has been made of the radiant energies emitted by coaxial flash lamps (or annular discharge) filled with rare gases at various pressures for current pulses of varying magnitude and duration. The energy inputs to the lamps ranged from 10-25 KJ, with quarter-cycle discharge times of less than one millisecond.

104. A Ruby Laser with an Elliptic Configuration. M. Ciftan, Proc. IRE, Vol. 49, pp. 960-961.

> A ruby laser with a new configuration was successfully operated at 1/15 input threshold energy of previous lasers. This decrease in pump energy was accomplished by a new configuration which efficiently focuses the pump radiation into the ruby rod.

105. On the Resonant Frequency Modes of Ruby Optical Masers. M. Ciftan, A. Krutchkoff, and S. Koozekanani, Proc. IRE, Vol. 50, pp. 84-85, January 1962.

> A laser beam is passed through a high-resolution spectrograph to obtain a detailed picture of the frequency modes.

106. A Ruby Laser with an Elliptic Configuration. M. Ciftan, C. F. Luck, C. G. Shafer, and H. Statz, Proc. IRE, Vol. 49, pp. 960-961, May 1961.

      The operation of a ruby laser with an elliptic pumping configuration at 1/15 the input threshold energy of previous lasers is reported.

107. High-Speed Photographic Study of the Coherent Radiation from a Ruby Laser. G. L. Clark, R. F. Wuerker, and C. M. York, J. Opt. Soc. Am., Vol. 52, pp. 878-880, August 1962.

      A high-speed image converter camera has been used to study the time variations as a function of position of the light emitted from pink ruby operated as a laser. The gross properties of the emitted light follow the theoretical predictions reasonably well, but the new experimental technique used in these studies reveals previously unreported details about the time variations of this light.

108. The Dynamic Pinch as a High-Intensity Light Source for Optical Maser Pumping. S. A. Colgate and A. W. Trivelpiece, pp. 288-292 in Advances in Quantum Electronics, J. R. Singer, ed., Columbia University Press, New York, 1961.

      See No. 109.

109. The Dynamic Pinch as a High-Intensity Light Source for Optical Maser Pumping. S. A. Colgate and A. W. Trivelpiece, University of California, Lawrence Radiation Laboratory, Livermore, California, UCRL-6364, March 1961.

      The optical radiation that occurs in a dynamic plasma pinch experiment depends in part on the amount and type of impurity atoms present. By proper choice of impurities, it should be possible to tailor optical radiation from the pinch in intensity, spectrum, and duration to provide good optical pumping for pulsed optical maser operation. This paper discusses the optical radiation and an experiment to use this radiation as the pump source for an optical maser.

110. Maser Oscillations in the Bouncing-Ball Modes of Large Resonators. R. J. Collins and J. A. Giordmaine. Bull. Am. Phys. Soc., II, Vol. 7, p. 446, August 1962.

Optical maser oscillations have been observed at discrete off-axis directions in a rectangular ruby rod having highly reflecting boundaries.

111. Modes of Optical Maser Oscillation in Closed Resonators. R. J. Collins and J. A. Giordmaine, Bell Telephone Laboratories, Murray Hill, New Jersey, Third International Symposium on Quantum Electronics, Paris, France, February 1963.

Optical maser emission has been studied in rectangular and circular cylindrical ruby rods having highly reflecting end surfaces, polished side walls, and consequently little mode selectivity. A rectangular rod with uniformly reflecting end surfaces fails to oscillate in the usual axial mode of circular rods; the emission is uniform in direction over a solid angle of at least 0.3 steradian.

112. Feedback Modulation with a Ruby Optical Maser. R. J. Collins and P. Kisliuk, Bull. Am. Phys. Soc., II, Vol. 6, p. 414, November 1961.

Modulation was accomplished through the construction of a laser with detached mirrors allowing a shutter to be placed between the ruby and one mirror. An increase in the output peak power was observed.

113. Control of Population Inversion in Pulsed Optical Masers by Feedback Modulation. R. J. Collins and P. Kisliuk, J. Appl. Physics, Vol. 33, pp. 2009–2011, June 1962.

The output power level of an optical maser is dependent on the level of inversion which can be reached. Using a technique in which the optical feedback is modulated by a shutter disk, an enhancement of the output power level of a pulsed ruby optical maser was observed. The higher level of output power occurred only for a short pulse. An explanation is given and the observed effect compared to the expected value.

114. Studies of the Emission from a Pulsed Ruby Optical Maser. R. J. Collins and D. F. Nelson, J. Opt. Soc. Am., Vol. 51, p. 473, April 1961.

The radiation output of a ruby optical maser has been studied under various conditions of temperature, reflectivity of end coating, and chromium content. Measurements have been made of the output power level, threshold pumping power, relaxation oscillations, and coherence.

115. Coherence, Narrowing, Directionality, and Relaxation Oscillations in the Light Emission from Ruby. R. J. Collins, D. F. Nelson, A. L. Schawlow, W. Bond, C. G. Garret, and W. Kaiser, Phys. Rev. Lett., Vol. 5, pp. 303-305, October 1960.

The observation of experimental effects proposed by Schawlow and Townes (in "Quantum Electronics" 1960) is described.

116. Interferometric Laser Mode Selector. S. A. Collins and G. R. White, Sperry Gyroscope Co., Great Neck, N. Y., Third International Symposium on Quantum Electronics, Paris, France, February 1963.

The mode selector consists of a laser cavity formed by external mirrors in which two small canted Fabry-Perot interferometers are interposed along with the active material. The interferometers serve as highly selective wavelength filters limiting the number of axial modes which will support oscillation. Theoretical design criteria are given. Experimental data are presented showing the effects of mode selection on wavelength, beam angle, threshold, and brightness.

117. Interferometer Laser Mode Selector. S. A. Collins and G. R. White, Appl. Optics, Vol. 2, pp. 448-449, April 1963.

The selector is composed of tilted Fabry-Perot etalons placed internal to the laser. It limits the spectrum radiated and narrows the beam angle.

118. Interferometer Laser Mode Selector. S. A. Collins and G. R. White, J. Opt. Soc. Am., Vol. 53, p. 514, April 1963.

Experiments are described which show beam angle narrowing in two dimensions as well as selection of axial frequencies, by the use of two internal mode selector etalons.

119. Absorption and Emission Frequencies in Lasers. W. J. Condell and H. I. Mandleberg, Appl. Optics,Supplement 1, pp. 84-85, 1962.

A possible means of mode selection is presented.

120. Investigation of Population Inversion in Helium. W. J. Condell, Jr., O. Van Gunten, and H. S. Bennett, J. Opt. Soc. Am., Vol. 50, pp.

184-185, February 1960.

It is concluded that the positive column of a discharge of pure helium will not afford optical maser action at pressures of 1.30 to 2.17 mm Hg and currents of 10-40 ma.

121. Comparison of Excitation Geometries for Lasers. R. S. Congleton, W. R. Sooy, D. R. Dewhirst, and L. D. Riley, Hughes Aircraft Co., Culver City, California, Third International Symposium on Quantum Electronics, Paris, France, February 1963.

Approximate relative efficiencies of various excitation geometries have been derived analytically and determined experimentally. Coaxial, elliptic, and close coupling in a minimum volume cavity is discussed.

122. Some Operating Characteristics of Flash-Pumped Ruby Lasers. J. C. Cook, Proc. IRE, Vol. 49, pp. 1570-1571, October 1961.

Operating characteristics of five lasers of different diameters, lengths, and C axis orientation are investigated.

123. Progress on Continuous Operation of Ruby Lasers. J. C. Cook, Bull. Am. Phys. Soc., II, Vol. 7, p. 118, February 1962.

Results of preliminary experiments in a pressurized arc-imaging furnace which produces a flux of 2400 W /cm^2 are reported. A water-cooled ruby has been operated as a laser for periods up to 0.5 second.

124. Output Power and Possible Continuous Operation of Ruby Lasers. J. C. Cook, W. L. Flowers, and C. B. Arnold, Proc. IRE, Vol. 50, pp. 330-331, March 1962.

The use of pink ruby in a high-power solid-state continuous-wave laser is discussed. Various methods for improving laser output are considered.

125. Measurement of Laser Output by Light Pressure. J. C. Cook, W. L. Flowers, and C. B. Arnold, Proc. IRE, Vol. 50, p. 1693, July 1962.

It is the purpose of this note to demonstrate that a practical system can be constructed to measure the pressure of a laser pulse on a reflecting surface and hence the total energy of the pulse.

126. Detection of Laser Radiation. V. J. Corcoran and Yoh-Han Pao, J. Opt. Soc. Am., Vol. 52, pp. 1341-1350, 1962.

> The magnitudes and frequencies of the fluctuations in electron emission from the photosurface associated with photon noise and shot noise and the parameters affecting these magnitudes are determined. Also, the influence of these fluctuations on the probability of detecting low-level radiation is investigated.

127. Experimental Transition Probabilities for Spectral Lines of Seventy Elements. C. H. Corliss and W. R. Bozman, NBS Mono. No. 53, 1961.

> The transition probabilities are listed.

128. The Electronic Spectra of Mixed Crystals. D. P. Craig and T. Thirunamachandran, Proc. Roy. Soc., A, Vol. 271, pp. 207-217, January 1963.

> The excited electronic states of dilute mixed crystals are discussed in terms of the theory of intermolecular interactions in dipole-dipole approximations. Resonance interactions of the Davydon type, which are of the first importance in pure crystals, are absent.

129. The Effects of Crystal Structure and Scattering Centers on the Light Pattern and Moding in Ruby Lasers. J. W. Crowe, J. Opt. Soc. Am., Vol. 53, p. 522, 1963.

> The pattern produced by laser oscillations in rubies is shown to be influenced by two types of crystal imperfections. Point imperfections nucleate the buildup of laser oscillations. Periodic striations produce optical paths which encourage the same patterns to build up under laser excitation.

130. A New Principle in the Design of a Millimetric Photo-Electric Laser Mixer. A. L. Cullen, University of Sheffield, Sheffield, U. K. Third International Symposium on Quantum Electronics, Paris, France, February 1963.

> In studying the design of a fast-wave photoelectric mixer for the generation of millimeter waves a simple method for controlling the phase velocity of the beat frequency wave was found. The beat phase velocity is found to depend in a simple

manner on the directions of the two laser beams in relation to the photoemissive surface. The dependence is worked out in detail, and examples relevant to the particular application are given.

131. A Proposed Fast-Wave Photo-Electric Laser Mixer for Millimeter Wave Generation. A. L. Cullen, Proc. Inst. Elec. Eng., Vol. 110, pp. 475–480, March 1963.

A photoelectric cell having the form of a parallel-strip transmission line in which the lower strip is photoemissive is proposed as a means of generating millimeter waves from two laser beams.

132. Proposals for Millimetric Photo-Mixing Using Surface Waves. A. L. Cullen, Polytechnic Institute of Brooklyn Symposia Series, XIII, Optical Masers, April 1963.

A fast-wave millimetric photo-mixer employing a parallel strip-line as the rf circuit has been proposed by the author. Two designs are proposed. In the first, the photoemissive coating is deposited on a corrugated surface capable of supporting a surface wave at the output frequency. Photoelectrons are collected by a suitably disposed anode. In the second, the anode is corrugated and can support a surface wave.

133. Zeeman Effects in Helium–Neon Lasers. W. Culshaw and J. Kannelaud, General Telephone and Electronics Laboratories, Palo Alto, California, Third International Symposium on Quantum Electronics, Paris, France, February 1963.

Relatively small magnetic fields are found to have a profound effect on the polarization and frequency content of He–Ne lasers. The experimental results due to the Zeeman effect are related to the theory.

134. Zeeman Effects in He–Ne Planar Laser. W. Culshaw, J. Kannelaud, and F. Lopez, Phys. Rev., Vol. 128, pp. 1747–1748, November 1962.

The effect of stray or applied magnetic fields on the modes and polarization of the planar laser output is considered.

135. Alkali Vapor Infrared Masers. H. Z. Cummins, I. Abella, O. S. Heavens, N. Kanable, and C. H. Townes, pp. 12–17 in Advances in

Quantum Electronics, J. R. Singer, ed., Columbia University Press, New York, 1961.

Experimental and theoretical details concerning the use of optically pumped cesium for maser action are discussed.

136. A High-Energy Laser Using a Multielliptical Cavity. H. Z. Cummins, Proc. IEEE, Vol. 51, pp. 254-255, January 1963.

The author concludes that the total power in a multielliptical cavity can never exceed the amount collected from a single source in an ideal single ellipse. Conflicting results are presented, in reply.

137. Fluorescence of the Trivalent Chromium Pair Spectrum in Synthetic Ruby. R. T. Daly, J. Opt. Soc. Am., Vol. 51, p. 473, April 1961.

The excitation spectrum of the strong ruby satellite lines at 7008 A and 7040 A have been explored at 77 and 2° K. The low-flying and excited levels of these lines appear to arise from two nonequivalent cation pairs in the corundum lattice. Results of experiments to produce stimulated emission in a three-level optical maser are given.

138. An Experimental Study of the Q-Switched Ruby Laser. R. T. Daly, TRG, Syosset, New York, Third International Symposium on Quantum Electronics, Paris, France, February 1963.

The time development of intense optical pulses was studied using both mechanically rotated reflectors and a Kerr-cell polarizer combination. Generation of single pulses containing up to 30% of the available stored energy and with a duration of 3 to 5 resonator time constants was possible with optimum combination of switching speed and prepumping. The transient behavior of a laser amplifier when the input is an intense pulse is also reported.

139. A Liquid Calorimeter for High-Energy Lasers. E. K. Damon and J. T. Flynn, Appl. Optics, Vol. 2, pp. 163-164, February 1963.

The calorimeter yields a temperature rise of about

10 °C for an input of up to 50 joules. The design of the calorimeter for higher ranges is investigated.

140. Photoionization of Gases by Optical Maser Radiation. Edward K.
Damon and Richard G. Tomlinson, Antenna Laboratory, Ohio State
University, Lasers and Applications Symposium, Ohio State Uni-
versity, November 1962.

Photoionization of noble and atmospheric gases in a focused
laser beam has been observed. The effect shows a strong
nonlinearity and appears to be power-dependent rather than
energy-dependent. The results of preliminary measurements using conventional and Q-switched laser pulses are
included.

141. Observation of Ionization of Gases by a Ruby Laser. E. K. Damon
and R. G. Tomlinson, Appl. Optics, Vol. 2, pp. 546-547, 1963.

Apparent photoionization due to 6943 A laser radiation has
been observed in recent experiments. By focusing coherent
radiation from a pulsed ruby laser into an evacuated cham-
ber, ionization has been produced in argon, helium, and a
neutral air mixture. This effect is a function of peak input
power and is strongly nonlinear. No detectable ionization
was produced by input pulses with peaks below 50 KW.

142. Semiclassical Treatment of the Optical Maser. L. W. Davis, Proc.
IEEE, Vol. 51, pp. 76-80, January 1963.

Using the semiclassical theory of radiation the steady-state
operation of the optical maser oscillator is studied in the
case where a single "cavity" mode is excited. On introducing certain simplifying assumptions, a straightforward
calculation leads to concise results for the frequency and
amplitude of the field oscillations. These results are either
well known or readily interpreted.

143. Interferometric Processing of a Phase-Modulated Optical Carrier.
W. F. Davison, Texas Instruments, Dallas, Texas, Third Inter-
national Symposium on Quantum Electronics, Paris, France,
February 1963.

With the optical maser there is now available radiation in
the optical region which approaches the spectral purity of
a carrier. The purpose of this paper is to emphasize the
necessity of accounting for the finite spectral width of the

optical carrier and to emphasize the utility of interferometric methods in optical communication by analyzing a specific phase–modulation − interferometric–demodulation scheme.

144. Electromagnetic Modes of an Optical Maser. E. S. Dayhoff, Bull. Am. Phys. Soc., II, Vol. 6, p. 365, June 1961.

A theory of wave propagation originally developed for treating the effects of diffraction in microwave interferometry is adapted to the discussion of modes of oscillation of an optical maser having plane–parallel ends.

145. Emission Patterns of a Ruby Laser. E. S. Dayhoff, Proc. IRE, Vol. 50, p. 1684, July 1962.

Results of an investigation of the filamentary mode patterns of a ruby laser with a high–speed camera are presented.

146. Laser Mode Studies in Solid Materials. E. S. Dayhoff, U. S. Naval Ordnance Laboratory, White Oak, Silver Spring, Maryland, Third International Symposium on Quantum Electronics, Paris, France, February 1963.

An idealized model of the transverse modes excited during individual spikes of light output in a solid–state laser is presented.

147. High–Speed Sequence Photography of a Ruby Laser. E. S. Dayhoff and B. Kessler, Appl. Optics, Vol. 1, pp. 339–341, May 1962.

The sequence of phenomena occurring when a ruby laser crystal is flashed is studied in a microsecond time scale by means of a high–speed framing camera at a rate of 500,000 frames per second.

148. Source Distribution in Ruby Lasers as a Function of Time. E. S. Dayhoff and B. V. Kessler, J. Opt. Soc. Am., Vol. 52, p. 594, May 1962.

This paper reports on studies of the distribution of light emission intensity across the radiating face of a ruby crystal during individual light pulses.

149. Ultrasonic Feedback Modulation of an Optical Maser Oscillator.

A. J. DeMaria and R. Gagosz, Proc. IRE, Vol. 50, p. 1522, June 1962.

The insertion of an ultrasonic cell between the reflecting end plates of a ruby optical maser has resulted in modulation of the optical feedback in the Fabry-Perot cavity. The ultrasonic feedback modulation is found to synchronize the usually random output pulses with the ultrasonic frequency. In addition, simultaneous increase in pulse heights and a decrease in pulse widths were observed.

150. Ultrasonic Refraction Shutter for Optical Maser Oscillators. A. J. DeMaria, R. Gagosz, and G. Barnard, J. Appl. Phys., Vol. 34, pp. 453-456, March 1963.

Experiments are described that demonstrate an ultrasonic shutter suitable for obtaining giant pulses from a ruby optical maser by utilizing the refractions resulting from the passage of a plane-parallel light beam through an ultrasonic field whose wavelength is much larger than the width of the light beam.

151. Ultrasonic Control of Laser Action. A. J. DeMaria, R. Gagosz, and G. Barnard, Polytechnic Institute of Brooklyn Symposia Series, XIII, Optical Masers, April 1963.

Time variations of the refractive index are obtained by propagating ultrasonic energy within the active Fabry-Perot cavity. Experiments are reported which demonstrate that the ultrasonic refraction effect can be used to synchronize the usually random output pulses of a ruby maser with the ultrasonic frequency. Experiments demonstrate that the ultrasonic refraction effect can be used as a shutter for the generation of single pulses of short duration and of extremely large amplitude.

152. X-Ray Excited Fluorescence in Crystalline Solids. V. E. Derr and J. J. Gallagher, Martin Marietta Corporation, Orlando, Florida, Third International Symposium on Quantum Electronics, Paris, France, February 1963.

The results of a measurement of the efficiency of the production of narrow-band light by x-ray excitation indicate that an x-ray or gamma-ray pumped ultraviolet laser may be feasible under pulsed operation, and the conditions necessary for oscillation will be discussed, with attention given to the problem of reduction of spontaneous emission at high frequencies.

153. Quantum Levels and Relaxation Time of Rare Earth Ions. O. Deutschbein and F. Auzel, C.N.E.T., Issy-les Moulineaux, Seine, France, Third International Symposium on Quantum Electronics, Paris, France, February 1963.

> Rare earth ions suitable for laser operation are studied in the form of pure salts and in the form of doping agents mixed in various solid supports having different crystal symmetry.

154. Composite Rod Optical Masers. G. E. Devlin, J. McKenna, A. D. May, and A. L. Schawlow, Appl. Optics, Vol. 1, pp. 11-15, January 1962.

> A new optical maser structure is described which reduces the threshold pumping power and increases the available power output. It consists of a composite rod which has a core of maser material covered by a coaxial sheath of transparent refractive material.

155. Microwave Generation in Ruby Due to Population Inversion Produced by Optical Absorption. D. P. Devor, I. J. D'Haenens, and C. K. Asawa, Phys. Rev. Lett., Vol. 8, p. 432, June 1962.

> Microwave amplification and generation by the stimulated emission of radiation were observed in ruby as a result of population inversion produced in the ground state of Cr^{3+} by the absorption of the coherent optical emission from a second ruby.

156. Laser Pumped Maser. D. P. Devor, I. J. D'Haenens, and C. K. Asawa, Hughes Research Laboratories, Malibu, California, Third International Symposium on Quantum Electronics, Paris, France, February 1963.

> Paramagnetic materials which demonstrate three-level laser operation may be used to obtain an optically pumped solid state maser. By using the same material for the active medium in both the laser and maser, the laser emission can be employed as the pump signal for the maser.

157. Stimulated Optical Emission in Ruby from 4.2°K to 300°K; Zero-

Field Splitting and Mode Structure. I. J. D'Haenens and C. K. Asawa,
Bull. Am. Phys. Soc., II, Vol. 6, p. 511, December, 1961.

The spectral character of the stimulated emission from ruby
has been investigated. It is found that zero field splitting
and cavity length determine the spectral character of the
emission. Mode structure is also discussed.

158. Stimulated and Fluorescent Optical Emission in Ruby from 4.2°K
to 300°K: Zero-Field Splitting and Mode Structure. I. J. D'Haenens
and C. K. Asawa, J. Appl. Phys., Vol. 33, pp. 3201-3208, November
1962.

The results of spectroscopic and stimulated emission experiments in ruby are described. Spectral character of the
stimulated emission was investigated. Zero-field splitting
of the $^4A_2$ ground level and the cavity optical length and its
reflectivity determine the spectral character of the emission.
At 300°K the system oscillates in a number of modes. At
4.2°K the system oscillates in off-axis modes unless the
length of the cavity is such that it has axial resonances co-
incident with the transition frequencies.

159. Stimulated and Fluorescent Optical Emission in Ruby from 4.2°K
to 300°K: Zero-Field Splitting and Mode Structure. I. J. D'Haenens
and C. K. Asawa, Hughes Research Laboratories, Malibu, California,
Third International Symposium on Quantum Electronics, Paris,
France, February 1963.

See No. 158.

160. Temperature and Concentration Effects in a Ruby Laser. I. J.
D'Haenens and V. Evtuhov, Hughes Research Laboratories, Malibu,
California, Third International Symposium on Quantum Electronics,
Paris, France, February 1963.

An analysis of the pumping requirement for a ruby laser as
a function of concentration, temperature, and pumping pulse
length has been made. This analysis predicts a minimum
in oscillation threshold as a function of concentration for a
given combination of laser parameters.

161. Coherence in Spontaneous Radiation Processes. R. H. Dicke, Phys.
Rev., Vol. 43, pp. 99-110, January 1954.

By considering a radiating gas as a single q-m system,

energy levels corresponding to certain correlations between individual molecules are described. Spontaneous emission of radiation in a transition between two such levels leads to the emission of coherent radiation.

162. Molecular Amplification and Generation Systems and Methods. R. H. Dicke, U. S. Patent 2,851,652, issued September 9, 1958.

> Proposes a number of arrangements and devices for energy storage with wide mode separation.

163. Optical Frequency Mixing in Bulk Semiconductors. M. DiDomentico, Jr., R. H. Pantell, O. Svelto, and J. N. Weaver, Appl. Phys. Letters, Vol. 1, pp. 77-79, December 1962.

> The results concern nonlinearities in bulk semiconductors, where the sample dimensions are two orders of magnitude larger than the i-layer in the p-i-n junction detector. Optical mixing has been obtained at frequencies as high as 10 times the cut-off frequency based on transit time effects.

164. The Spectroscopy of Trivalent Rare Earth Ions. G. H. Dieke, Polytechnic Institute of Brooklyn Symposia Series, XIII, Optical Masers, April 1963.

> Illustrations are presented which show that optical spectroscopy can accurately establish the energy levels of rare earth ions in crystals. The energy transfer between the ions and the lattice, and between different ions, can be established by observation of the monochromatically excited fluorescence spectrum.

165. Spectroscopic Observations on Maser Materials. G. H. Dieke, pp. 164-186 in Advances in Quantum Electronics, J. R. Singer, ed., Columbia University Press, New York, 1961.

> Experimental methods, determination of stabilities of energy levels, influence of concentration on energy levels, magnetic properties, and line width are topics covered in this tutorial paper.

166. Confocal Resonators with Periodic Reflectivity. G. T. DiFrancia, Polytechnic Institute of Brooklyn Symposia Series, XIII, Optical Masers, April 1963.

> All the optical resonators consisting of two, real or virtual,

plane mirrors located at the two focal planes of an optical system have the same properties as the confocal resonator described by Fox and Li and Boyd and Gordon. The field at one mirror is the Fourier transform of the field at the other, and vice versa. By this principle one can derive a great variety of optical systems which are all equivalent from the point of view of mode analysis, while having different geometries. The physical insight gained by these considerations can be of great help in assessing the influence of aberrations and in finding approximate solutions in complicated situations.

167. Chromium Distribution in Synthetic Ruby Crystals. R. R. Dils, G. W. Martin, and R. A. Huggins, Appl. Phys. Lett., Vol. 1, pp. 75-76, December 1962.

Microscopic heterogeneity of considerable magnitude and on a fine scale was found in five ruby crystals. The overall chromium content of the crystals varied from 0.03 to 0.5%.

168. A Theory of Pumping by Incoherent Waves. H. Dormont, Laboratories d'Electronique et de Physique Appliquée, Paris, Third International Symposium on Quantum Electronics, Paris, France, February 1963.

The behavior of a three-level atomic system within an interferometer is studied with the help of quantum field theory.

169. Influence of Finite Bandwidth on the Interaction of Light Waves in a Nonlinear Dielectric. J. Ducuing and J. A. Armstrong, Gordon McKay Laboratory, Harvard University, Third International Symposium on Quantum Electronics, Paris, France, February 1963.

A formalism has been developed for treating the effect of a small finite bandwidth on the interaction of light waves in a nonlinear medium. Particular attention has been given to second harmonic generation. In the case where second harmonic generation has resulted in appreciable depletion of the power at the fundamental, it is found that the relative bandwidth of the second harmonic does not increase greatly but is only slightly larger than the initial relative bandwidth of the fundamental.

170. Propagation of Three Optical Plane Waves in a Nonlinear Dielectric Medium. J. Ducuing, J. Armstrong, and N. Bloembergen, Bull. Am. Phys. Soc., II, Vol. 7, p. 196, March 1962.

The treatment of nonlinear coupling in a dielectric medium is extended to the case of three plane waves which satisfy the energy and momentum relations.

171. Spontaneous Radiative Recombination in Semiconductors. W. P. Dumke, Phys. Rev., Vol. 105, pp. 139-144, January 1957.

> The mechanisms by which electrons and holes recombine with the emission of radiation are examined.

172. Interband Transitions and Maser Action. W. P. Dumke, Phys. Rev., Vol. 127, pp. 1559-1563, September 1962.

> The possibility of using interband transitions to achieve maser action is considered. The criterion for maser action is presented in a way which allows the most direct use of optical absorption data. The absorption constant for interband transitions is related to the normal absorption constant for direct and indirect exciton transitions.

173. Electromagnetic Mode Population in Light-Emitting Junctions. W. P. Dumke, IBM J. Res. Dev., Vol. 7, pp. 66-67, January 1963.

> A calculation of the electromagnetic mode population in a light-emitting crystal is presented. It is shown that such modes at high injection currents should be highly excited.

174. Properties and Mechanisms of GaAs Injection Lasers. W. P. Dumke, Polytechnic Institute of Brooklyn Symposia Series, XIII, Optical Masers, April 1963.

> The structure and some properties of GaAs injection lasers are described. The thresholds for lasing and the efficiencies of these devices are related to the absorption and recombination mechanisms in GaAs.

175. Coherent Light, Its Production, Its Applications. C. Dumousear, Rev. Soc. Roy. Belge. Ingen. Industr., No. 3, pp. 154-162, March 1962.

> The principles of coherent transmission, the physics of maser operation, and the applications of optical masers are qualitatively discussed.

176. Direct Observation of Longitudinal Modes in the Outputs of Optical

Masers. R. C. Duncan, Jr., Z. J. Kiss, and J. P. Wittke, J. Appl. Phys., Vol. 33, pp. 2568-2569, August 1962.

Longitudinal modes in the output of pulsed $U:CaF_2$ and ruby optical masers have been observed directly by means of a high-resolution, high-dispersion grating spectrometer. The mode spacings are in reasonable agreement with the expected values. The line width of the individual modes is instrument-limited at about 0.05 cm^{-1}. Maser action is observed to occur simultaneously in the various modes.

177. Theory of Relaxation Oscillations in Optical Masers. R. Dunsmuir, J. Electronics and Control, Vol. 10, pp. 453-458, June 1961.

A theoretical explanation of the spikes present in the light output of optical masers is presented. Simple formulae are given for the time interval between spikes and the rate at which they decay to a constant output level.

178. The Use of Interference Effects to Improve the Efficiency of Optical Modulators. C. C. Eaglesfield and M. M. Ramsay, Standard Tele-communication Laboratories Limited, Harlow, Essex, U. K., Third International Symposium on Quantum Electronics, Paris, France, February 1963.

Considerable improvement in efficiency can be effected by reducing the volume of the active element, and further im-provements can be obtained by the use of interference ef-fects. A system is described in which the active element forms the coupling between a microwave and an optical res-onator.

179. Double Pulse Excitation of a Ruby Laser. J. L. Emmett and R. W. Hellwarth, Bull. Am. Phys. Soc., II, Vol. 7, p. 615, December 1962.

The ruby is initially pumped in the usual manner. Shortly after threshold is reached it is pumped again with a higher power pulse, increasing the peak pump power by a factor of 50. Results are presented and compared with the predic-tions of various theories.

180. Observations Relating to the Transverse and Longitudinal Modes of a Ruby Laser. V. Evtuhov and J. K. Neeland, Appl. Optics, Vol. 1, pp. 517-520, July 1962.

The mode characteristics of an optical resonator are described. By operating a ruby laser sufficiently close to oscillation threshold and using a high-quality crystal, the variations of the beam intensity across the partially transmitting face of the laser crystal have been directly observed.

181. Characteristics of Ruby Laser Modes in a Plane-Parallel Resonator. V. Evtuhov and J. K. Neeland, Hughes Research Laboratories, Malibu, California, Third International Symposium on Quantum Electronics, Paris, France, February 1963.

Experimental data describing the behavior of longitudinal modes, near-field patterns, and the emitted laser beam are summarized. An attempt is made to establish possible reasons for the observed behavior.

182. Measurements and Interpretation of Laser Beam Divergence. V. Evtuhov and J. K. Neeland, Appl. Optics, Vol. 2, pp. 319-320, March 1963.

A sapphire-clad ruby 3.23 cm long and 2 mm in diameter with plane-parallel faces was used in the experiment. Within the accuracy of the measurements the beams start out parallel to each other and then begin to diverge; in the far field they diverge at a constant angle of approximately 2.14 mrad.

183. Laser Patent. V. A. Fabrikant, M. M. Vadynski, and F. A. Butayeva, AD 287288, 20 pp., October 1962.

A verbatim translation of a patent which appeared in the July 1962 issue of Zhurnal Izobreteniy, the official Soviet patent organ, is given. A method of amplifying electromagnetic radiation based on a medium with negative coefficient of absorption and distinguished by the application of a multiple passage of the signal through the amplifying medium in order to increase gain is described.

184. Self-Excitation Conditions of a Laser. V. M. Fain and Ya. I. Khanin, Zhur. Eksptl. Teoret. Fiz., Vol. 41, pp. 1498-1502, 1961 (in Russian): Soviet Physics-JETP, Vol. 14, p. 1069, 1962.

Equations are derived describing the self-excitation in a molecular generator with a cavity having demensions greater than the generated wavelength.

185. Gas Optical Maser Operating at Wavelengths between 2 and 9

Microns. W. L. Faust, R. A. McFarlane, C. K. Patel, and G. G. Barrett, Bull. Am. Phys. Soc., II, Vol. 7, p. 553, November 1962.

> A gas optical maser for general use in study of the stimulated emission spectra of the noble gases beyond 2.2 microns is described. Oscillation has been obtained on 18 lines.

186. Semiconductor Lasers: A Brilliant New Source of Light. D. Fishlock, New Scientist, Vol. 17, pp. 65-67, January 1963.

> The development of the semiconductor laser in the United States and Great Britain is reported.

187. Research and Investigation of Materials for Laser Applications. R. Fitzpatrick and S. E. Sobottka, AD286216, 10 pp., September 1962.

> Research being conducted to prepare and optimize semiconductor materials for electron injection laser applications is described.

188. Research and Investigation of Materials for Laser Applications. R. Fitzpatrick and S. E. Sobottka, AD294771, 15 pp., December 1962.

> Techniques for developing germanium arsenide diodes have been developed. The effect of doping with neodymium is investigated.

189. Effects of γ-Irradiation on the Performance of a Ruby Laser. W. Flowers and J. Jenney, Proc. IEEE, Vol. 51, pp. 858-859, 1963.

> A significant increase in the efficiency of ruby laser rods was obtained by irradiating the rods with Co^{60} gamma rays. This is due to a more efficient absorption of the pump light by color centers produced in the ruby by irradiation.

190. Plasma-Impingement Mechanism for High-Power Laser Pumping. G. Fonda-Bonardi, Litton Systems, Inc., Lasers and Applications Symposium, Ohio State University, November 1962.

> Conventional pumping utilizes the light generated in an electrical discharge in a flash tube. Experiments were carried out which showed an extremely intense output of light from the area of impingement of a fast-moving (10^5 meters/second) plasmoid on a solid surface. Accordingly, further

experiments were carried out to utilize this mechanism for the generation of the pumping light. The electrical discharge occurs in a gas completely surrounding the laser crystal, and the geometry of the system is arranged to obtain a concentric snowplow effect, whereby a large amount of electrical energy is converted into kinetic energy of the plasma, and this is directed toward the laser crystal. The plasma is caused to impinge on the outer surface of the crystal, where much of the kinetic energy is reconverted to light energy.

191. Scanning Fabry-Perot Observation of Optical Maser Output. R. L. Fork, E. I. Gordon, D. R. Herriot, H. W. Kogelnik, and J. W. Loofbourrow, Bull. Am. Phys. Soc., II, Vol. 8, p. 380, April 1963.

Direct observation of the mode structure of optical masers operated in both the visible and infrared at various gains has been carried out by means of scanning Fabry-Perot interferometers of several designs.

192. "Negative" Tensor Susceptibility and Application to Light Modulation. R. L. Fork and C. K. Patel, Bull. Am. Phys. Soc., II, Vol. 7, p. 615, December 1962.

A theoretical analysis of the "negative" dielectric susceptibility tensor for a medium in which some of the levels exhibit population inversion has been made. The analysis shows that the medium will exhibit a large "negative" birefringence and dichroism in the vicinity of a resonance occurring between levels with inverted populations. It is shown that these effects are applicable to modulation of coherent light.

193. Photoelectric Mixing as a Spectroscopic Tool. A. T. Forrester, J. Opt. Soc. Am., Vol. 51, pp. 253-259, March 1961.

Photoelectric mixing when combined with lasers may provide the basis of relatively simple optical measurements using radio-frequency-like receivers. Two receiver types are considered. The low-level receiver offers the possibility of simple observations of the special shape of the laser output. The superheterodyne receiver offers the possibility of measurement of the shapes of ordinary spectral lines.

194. Photodetection and Photomixing of Laser Outputs. A. T. Forrester,

pp. 233-238 in Advances in Quantum Electronics, J. R. Singer, ed.,
Columbia University Press, New York, 1961.

> Low-level detection of laser signals is considered. The advantages of using superheterodyne receivers are pointed out.

195. Resonant Modes in an Optical Maser. A. G. Fox and T. Li, Proc.
IRE, Vol. 48, pp. 1904-1905, November 1960.

> The experiment of bouncing a wave back and forth between a zero-reflection-loss plane-parallel mirrors was programmed on a computer. The results indicate many normal modes representing higher orders of variation of field across the end plates.

196. Resonant Modes in a Maser Interferometer. A. G. Fox and T. Li,
pp. 308-317 in Advances in Quantum Electronics, J. R. Singer, ed.,
Columbia University Press, New York, 1961; Bell System Technical Journal, Vol. 40, pp. 453-488, March 1961.

> A theoretical investigation of diffraction of electromagnetic waves in Fabry-Perot interferometers when used as resonators in optical masers reveals that a steady state is reached in which the relative field distribution does not vary from transit to transit and the amplitude of the field decays at an exponential rate. Many such steady states or normal modes are possible depending on initial wave distribution. An electronic digital computer was programmed to compute the electromagnetic field across the mirrors of the interferometer where an initially launched wave travels back and forth between two mirrors.

197. Modes in a Maser Interferometer with Curved and Tilted Mirrors.
A. G. Fox and T. Li, Proc. IEEE, Vol. 51, pp. 80-89, January 1962.

> Results of a study of the effects of certain simple forms of aberration in the Fabry-Perot interferometer are presented. The first is represented by tilted plane mirrors and the second by curved mirrors. Tilting causes asymmetric mode patterns and greater diffraction losses, as well as a beating phenomenon due to equalization of losses of the two lowest order modes. Curved mirrors cause regions of low and high loss as mirror spacing is varied.

198. Modes in a Maser Interferometer with Curved Mirrors. A. G. Fox
and T. Li, Bell Telephone Laboratories, Holmdel, N. J., Third

International Symposium on Quantum Electronics, Paris, France, February 1963.

An interferometer consisting of mirrors of arbitrary curvature is studied. Mode patterns and their diffraction losses have been computed for the high- and low-loss regions due to mirror spacing.

199. On Diffraction Losses in Laser Interferometers. A. G. Fox , T. Li, and S. P. Morgan, Appl. Optics, Vol. 2, pp. 544-545, 1963.

Tang has calculated, using the variational principle, the diffraction losses of the modes in an optical maser interferometer. The authors comment briefly on the paper by Tang.

200. Power and Efficiency Considerations in Continuous Laser Operation. D. R. Frankl, J. Appl. Phys., Vol. 34, pp. 459-462, March 1963.

Various factors entering into the continuous operation of optically pumped solid-state lasers are summarized. Numerical estimates for ruby and for Nd^{3+}-activated materials in two types of optical systems suggest that several watts of output power should be obtainable when pumping with a 1-KW mercury arc lamp.

201. Generation of Optical Harmonics. P. A. Franken, A. E. Hill, C.W. Peters, and G. Weinreich, Phys. Rev. Lett. , Vol. 7, pp. 118-119, August 1961.

Second harmonics are observed in an intense beam of 6943 A light projected through crystalline quartz.

202. High-Energy Experiments with Optical Masers. P. A. Franken, J. Opt. Soc. Am., Vol. 52, p. 601, May 1962.

The extraordinary intensity of ruby lasers permits brief pulses in excess of 100 mw/cm^2 at the focal planes of simple optical systems. Experiments that utilize these high powers are reviewed.

203. Efficiency of a Multiple-Ellipse Confocal Laser Pumping Configuration. D. L. Fried and P. Eltgroth, Proc. IRE, Vol. 50, p. 2849, December 1962.

The efficiency of multielliptic cavities is shown to be very little higher than that for a single-ellipse cavity.

204. Output Power Oscillations in a Ruby Laser. O. G. Fritz, Jr., and
K. Tokunaga, Bull. Am. Phys. Soc., II, Vol. 7, p. 397, June 1962.

> A pair of coupled equations are obtained by an application of
> the three-level rate equations for ruby. Under the assump-
> tion that the largest loss factor is associated with the reflec-
> tivity of the ends of the laser, the variation of oscillation
> frequency with the reflectivity of the ends of the laser is ex-
> amined.

205. Photomultiplication with Microwave Response. O. L. Gaddy and
D. F. Holshouser, University of Illinois, Urbana, Illinois, Third
International Symposium on Quantum Electronics, Paris, France,
February 1963.

> The principles of a method of which secondary emission multi-
> plication are described in which current gain approaching
> that of conventional electrostatic photomultipliers is possible
> with the spread in electron transit time being restricted to
> a small part of a microwave period. This is accomplished
> by using microwave electric fields as a source of energy for
> secondary emission combined with a steady magnetic field.
> Experiments are described in which light modulated at 3 Gc/s
> is detected with current gain of 10^5.

206. Coherence and Beam Directivity of an Optical Ruby Oscillator.
M. D. Galanin, A. M. Leontovich, and Z. A. Chizikovu, AD286350,
23 pp., October 1962.

> The relationship between coherence and beam directivity of
> a ruby laser was studied. It was shown that radiation pulses
> during emission occur simultaneously over the entire radi-
> ating area of the crystal. It was shown that emission pulses
> always occur simultaneously in various regions of the cry-
> stal, though at times there may be differences in the rela-
> tive peak size.

207. On the Intensity Interferometer with Coherent Background. Hildeya
Gamo, pp. 252-266 in Advances in Quantum Electronics, J. R.
Singer, ed., Columbia University Press, New York, 1961.

> Two types of modified intensity interferometers are obtained
> by superposing a highly monochromatic spatially coherent

background upon an incident beam. The signal-to-noise ratio in these interferometers is generally improved by increasing the intensity of the coherent background.

208. Generation and Radiation of Ultra Microwaves by Optical Mixing. O. P. Gandhi, Proc. IRE, Vol. 50, pp. 1829-1830, August 1962.

A scheme of nonlinear mixing is described which provides a simultaneous generator and highly directional radiator of ultramicrowaves.

209. Stimulated Emission of Ultraviolet Radiation from Gadolinium-Activated Glass. H. W. Gandy and R. J. Ginther, Appl. Phys. Lett., Vol. 1, pp. 25-27, September 1962.

Stimulation emission of radiation has been observed from gadolinium-activated lithium-magnesium aluminosilicate glass cooled to liquid nitrogen temperature at a wavelength of 3125 A.

210. Stimulated Emission from Holmium-Activated Glass. H. W. Gandy and R. J. Ginther, Proc. IRE, Vol. 50, pp. 2113-2114, October 1962.

Stimulated emission of radiation has been observed in a holmium-activated $LiMgAlSiO_3$ glass at liquid nitrogen temperature.

211. Simultaneous Laser Action of Neodymium and Ytterbium Ions in Silicate Glass. H. W. Gandy and R. J. Ginther, Proc. IRE, Vol. 50, pp. 2114-2118, October 1962.

Stimulated emission has been observed simultaneously from two different ions contained in the same $LiMgAlSiO_3$ glass etalon while being operated at liquid nitrogen temperature.

212. Stimulated Emission of Ultraviolet Radiation from Gadolinium-Activated Glass. H. W. Gandy and R. J. Ginther, U. S. Naval Research Laboratory, Washington, D. C., Third International Symposium on Quantum Electronics, Paris, France, February 1963. Appl. Phys. Lett.,Vol. 1, pp. 25-27, 1962.

Stimulated emission of radiation has been observed from gadolinium-activated lithium-magnesium aluminosilicate glass cooled to liquid nitrogen temperature at a wavelength of 3125 A.

213. Infrared Gas Optical Masers. C. G. Garrett, Polytechnic Institute of Brooklyn Symposia Series, XIII, Optical Masers, April 1963.

Recent work on the noble gas optical masers holds out considerable promise for closing the gap between the optical and microwave portions of the spectrum. Oscillation has been achieved so far on around 150 different transitions. The longest wavelength achieved so far is about 28 microns (in neon).

214. Monochromaticity and Directionality of Coherent Light from Ruby. C. G. Garrett, W. L. Bond, and W. K. Kaiser, Bull. Am. Phys. Soc., II, Vol. 6, p. 68, February 1961.

Line sharpening, spiking, and a mosaic structure observed in the output of the ruby maser are reported.

215. Fluorescence and Optical Maser Effects in Rare-Earth-Doped Calcium Fluoride. C. G. Garrett and W. Kaiser, J. Opt. Soc. Am., Vol. 51, p. 477, April 1961.

Measurements of emission spectra, absorption spectra, absolute activation spectra, and fluorescent decay lifetime have been made on samples of rare-earth-doped calcium fluorides with a view to finding suitable materials for optically pumped solid-state optical masers.

216. Stimulated Emission into Optical Whispering Modes of Spheres. C. G. Garrett, W. Kaiser, and W. L. Bond. Phys. Rev.,Vol. 124, pp. 1807-1809, December 1961.

Stimulated emission into optical whispering modes of a spherical sample of $CaF_2 : Sm^{++}$ has been observed. Light produced by the stimulated emission is radiated tangentially from each point on the surface of the sphere.

217. Fluorescence and Optical Maser Effects in Calcium Fluoride : Divalent Samarium. C. G. Garrett, W. Kaiser, and D. L. Wood, pp. 77-78, in Advances in Quantum Electronics, J. R. Singer, ed., Columbia University Press, New York, 1961.

Results of an investigation of the absorption and emission spectrum of calcium fluoride:divalent samarium crystals are reported. Calcium fluoride crystals doped with varying concentrations of samarium are used, and maser action was investigated over a large range of pumping powers.

218. Laboratory Alkali Metal Vapor Lamps for Optical Pumping Experiments. V. B. Gerard, J. Sci. Instr., Vol. 39, pp. 217-218, 1962.

The construction of electrodeless calcium and rubidium lamps with high and constant light output of the D lines, low noise and low self-reversal is described.

219. Optical Lattice Filters from the Wave Field of Laser Radiation. R. Gerhatz, Proc. IEEE, Vol. 51, pp. 862-863, 1963.

The advance of laser techniques has offered a chance for producing periodic layers in transparent solids by direct action of the monochromatic beam. This is accomplished by the field strength of the light from a laser, which is many orders of magnitude higher than that from ordinary intense sources.

220. Recent Developments in Maser Devices and Materials. H. J. Gerritsen, Appl. Optics, Vol. 1, pp. 37-44, January 1962.

The properties of crystals used in solid-state masers are discussed. A table is presented summarizing this information for present maser materials. It is pointed out that further material research needs to be done to improve the performance of masers and extend their frequency range.

221. Optical Detection of Paramagnetic Resonance in an Excited State of Cr^{3+} in Al_2O_3. S. Geschwind, R. J. Colling, and A. L. Schawlow, Phys. Rev. Lett., Vol. 3, pp. 545-548, December 1959.

The method of optical detection makes use of the selective reabsorption in the ground-state Zeeman levels of the fluorescent light from excited states in solids at very low temperatures.

222. Sharp-Line Fluorescence, Electron Paramagnetic Resonance, and Thermoluminescence of Mn^{4+} in Al_2O_3. S. Geschwind, P. Kisliuk, M. P. Klein, J. P. Remeika, and D. L. Wood, Phys. Rev., Vol. 126, pp. 1684-1686, June 1962.

Sharp-line fluorescence, paramagnetic resonance in the

ground state, and optical absorption due to Mn^{4+} in alpha Al_2O_3 have been observed. The possible application to light masers is briefly discussed.

223. A Unidirectional Traveling-Wave Optical Maser. J. E. Geusic and H. E. Scovil, B.S.T.J., Vol. 41, pp. 1371-1397, July 1962.

The basic ideas leading to a unidirectional traveling-wave optical maser are presented. Experimental data on the performance of pulsed ruby amplifying sections and high-density PbO glass Faraday rotation isolators are given. Feasibility tests on a two-section device have been made and are in agreement with predictions. Some remarks are made concerning image definition, channel capacity, noise, and pump power requirements.

224. Mixing of Light Beams in Crystals. J. A. Giordmaine, Phys. Rev. Lett., Vol. 8, pp. 19-20, January 1962.

The mixing of plane light waves having different directions of propagation and the attainment of coherence volumes of about 0.2 cm^3 in the production of second harmonic radiation in potassium dihydrogen phosphate KDP are discussed.

225. Measurement of the Absorption Spectrum of Ruby Excited for the Study of Its Operation in an Optical Maser. F. Gires and G. Mayer, Compt. rend. acad. sci. (France), Vol. 254, pp. 659-661, 1962 (in French).

It is shown experimentally that irradiation with a powerful light source strongly depopulates the ground state energy level $4H_2$ for chromium in ruby, resulting in increased population of the ^2E levels. Normal absorption bands are weakened and new bands due to transitions from levels ^2E appear. These effects allow the population of the $4H_2$ level to be determined.

226. Optical Attenuation and Amplification in Strongly Excited Ruby. F. Gires and G. Mayer, Compagnie Générale de Telegraphie Sans Fil, Paris, France, Third International Symposium on Quantum Electronics, Paris, France, February 1963.

The optical transmission of ruby has been measured as a function of wavelength and polarization in various conditions of optical excitation.

227. A Method for Calibration of Laser Energy Output. A. L. Glick, Proc. IRE, Vol. 50, p. 1835, August 1962.

> The laser energy output is calibrated by attenuating the laser beam with neutral filters, directing it into a phototube and integrating the current produced. This charge is directly dependent on the number of light quanta and therefore the light energy.

228. Application of Total Internal Reflection Prisms for Gaseous Lasers. Z. Godzinski, Proc. IEEE, Vol. 51, p. 361, February 1963.

> The difficulty of very small mirror alignment tolerances in the construction of gaseous lasers can be avoided by using total internal reflection corner prisms instead of mirrors.

229. Notes on Coherence vs. Narrow-Bandedness in Regenerative Oscillators, Masers, Lasers, Etc., M. J. Golay, Proc. IRE, Vol. 49, pp. 958-959, May 1961.

> The distinction between narrow-bandedness and coherence is reviewed. A test for coherence is suggested.

230. Pathology of the Effect of the Laser Beam on the Skin. L. Goldman, D. L. Blaney, D. J. Kindel, Jr., D. Richfield, and E. K. Frank, Nature, Vol. 197, pp. 912-914, March 1963.

> Preliminary investigations with a lightweight solid laser with a ruby crystal showed that superficial destructive lesions may be produced in the skin of man. The more intense the coloring of the skin the deeper the reaction.

231. A Laser Design for Space Communications. L. Goldmuntz, IRE Int. Conv. Rec., Part 5, Vol. 10, pp. 298-305, March 1962.

> Performance characteristics for various missions for simple laser systems utilizing both coherent and incoherent detection are considered.

232. An Analysis of General Optical Resonator Systems. D. Golge, Institut für Höchstfrequenztechnik, Technische Hochscule Braunschweig, Germany, Third International Symposium on Quantum Electronics, Paris, France, February 1963.

> Since the modes of the confocal resonator are highly degen-

erate, care must be taken in applying the theory developed by Boyd and Gordon to nondegenerate systems. The general analysis for mirror systems of any curvature and distance given here leads to integral equations with integrals of the type of a finite Gauss transform, containing the finite Fourier transform for the special case of confocal mirrors.

233. The Fabry-Perot Electro-Optic Modulator. E. I. Gordon and J. D. Rigden, B.S.T.J., Vol. 42, pp. 155-180, January 1963.

    The modulator consisting of Fabry-Perot etalon plates separated by an electro-optic material such as KDP is analyzed in detail.

234. Interactions and Saturation in He-Ne Maser Levels. E. I. Gordon, A. D. White, and J. D. Rigden, Polytechnic Institute of Brooklyn Symposia Series, XIII, Optical Masers, April 1963.

    In the He-Ne maser there are at least three significant transitions which have levels in common. Experiments are described which have used the competing maser transitions to determine some of the physical constants of the He-Ne system. An expression for the power output of an inhomogeneously broadened maser transition will be derived which accurately predicts the output power under all conditions of gain per unit length and reflection loss.

235. Coherent Optical Emission from Molecular Beams. I. Gorog, University of California, Berkeley, Institute of Engineering Research, Ser. 60, issue 418, N62-12154, November 1961.

    The feasibility of constructing an optical maser where the excited atoms are obtained as products of photodissociation of alkali halides is considered. It is shown that by confining the active particles to a molecular beam, an excess reduction in the output line width can be obtained. Calculations show that maser oscillation in such a system is possible.

236. Laser Wavelength and Frequency Standard. G. Gould, J. Opt. Soc. Am., Vol. 53, p. 515, April 1963.

    Methods for centering a cw gas laser oscillator frequency on the fluorescent line have been studied theoretically and experimentally. A piezoelectric crystal is used to control cavity length and hence the oscillator frequency.

237. Light Sources for Optical Pumping. J. P. Gourber, Compagnie Générale de Télégraphie Sans Fil, Orsay, France, Third International Symposium on Quantum Electronics, Paris, France, February 1963.

> Experiments indicate that to provide a highly stable and noise-free gaseous discharge in an alkali metal vapor it is desirable to eliminate the use of direct or low-frequency voltage and also to determine the optimum frequency and power level of the excitation, identify the respective influences of the electrical and magnetic fields, define the temperature and thermal gradient in the vicinity of the light source, and use a buffer gas (krypton) in order to facilitate the ignition.

238. A Technique for Obtaining Single, High-Peak-Power Pulses from a Ruby Laser. D. G. Grant, Proc. IEEE, Vol. 51, p. 604, April 1963.

> A method is described which employs the use of an aluminized mylar film. Switching is accomplished through the burning of a hole in the film by the laser beam.

239. A Correction to the Townes' Line Width Formula in the Nonlinear Domain. P. A. Grivet and A. Blaquiere, Polytechnic Institute of Brooklyn Symposia Series, XIII, Optical Masers, April 1963.

> A maser or laser oscillator is very similar to a classical feedback oscillator even in the nonlinear domain. Taking account of nonlinearities indicates how to define accurately conditions of validity for the analogy; one then obtains a formula identical to Townes' formula, but with a corrective factor of one half.

240. Enhancement in Mercury-Krypton and Xenon-Krypton Gaseous Discharges. G. Grosof and R. Targ, Appl. Optics, Vol. 2, pp. 299-304, March 1963.

> Spectroscopic examination of electrodeless gaseous discharges in krypton containing small amounts of xenon or mercury shows a directed transfer from the krypton 1s5 metastable atom to the 2s levels in xenon and the 9 1pl state in mercury.

241. Study of the Amplification of a Gas Mixture Using a Laser. R. Grudzinski and J. Spalter, Compagnie Générale d'Electricité Marcoussis, S.-et-O., France, Third International Symposium on

Quantum Electronics, Paris, France, February 1963.

> A He-Ne laser is used as a light source for a precise measurement of the amplification of a He-Ne mixture. The test tube is excited with a dc discharge.

242. Beats and Modulation in Optical Ruby Masers. K. Gurs, Siemens and Halske A. G., Forschungslaboratorium, Munich, Germany, Third International Symposium on Quantum Electronics, Paris, France, February 1963.

> In a proper setup the ruby maser exhibits a component of continuous emission during the pumping pulse, after damping of the spikes. On changing the operating conditions, periodic changes of power output are observed, which can be explained as beats between adjacent axial modes. Small changes in the Q of the cavity with the beat frequency secure full modulation of the emission.

243. Internal Modulation of Optical Masers. K. Gurs and R. Muller, Polytechnic Institute of Brooklyn Symposia Series, XIII, Optical Masers, April 1963.

> The insertion of an electrical birefringent material into the feedback path of an optical maser provides the full modulation of the emitter light by only a small rotation of the plane of polarization within the modulating material. The dynamic behavior is described. The conditions for microwave modulation and the usable bandwidth are discussed. Experimental results are presented.

244. Coherent Light Emission from GaAs Junctions. R. N. Hall, G. E. Fenner, J. D. Kingsley, T. J. Soltys, and R. D. Carlson, Phys. Rev. Lett., Vol. 9, pp. 366-368, November 1962.

> The first instance of direct conversion of electrical energy to coherent infrared radiation in a solid-state device is reported. The laser is the first to involve transitions between energy bands rather than localized atomic levels.

245. FM-AM Optical Converter. S. E. Harris, Stanford Electronics Laboratories, Stanford University, Stanford California. Third International Symposium on Quantum Electronics, Paris, France, February 1963.

A simple and practical device similar to a single-state Lyot filter has been proposed and experimentally demonstrated for the conversion of microwave frequency-modulated light to microwave amplitude-modulated light. The device appears to be stable, easily controlled, insensitive to vibration, and essentially lossless. It requires well-collimated light.

246. Modulation and Direct Demodulation of Coherent and Incoherent Light at a Microwave Frequency. S. E. Harris, B. J. McMurtry, and A. E. Siegman, Appl. Phys. Lett., Vol. 1, pp. 37-39, October 1962.

The modulation and direct demodulation of optical radiation using a cavity-type KDP Pockels cell and a microwave light detector together with both the coherent light from a pulsed ruby laser and the incoherent light from a mercury arc are reported.

247. Proposed Microwave Phototube for Demodulating FM Light Signals. S. E. Harris and A. E. Siegman, Stanford Electronics Laboratories Tech. Report 176-1, March 1962.

A microwave phototube for demodulating frequency-modulated light signals is proposed. The demodulation is based upon the conversion of the FM light to space-modulated light via an optical dispersing element. This space-modulated light is then incident on a photocathode, where it is the source of transverse electron beam waves.

248. A Proposed FM Phototube for Demodulating Microwave-Frequency Modulated Light Signals. S. E. Harris and A. E. Siegman, IRE Trans. on Electron Devices, Vol. ED-9, pp. 322-328, July 1962.

A microwave phototube for demodulating frequency-modulated light signals is proposed.

249. Optical Heterodyning and Optical Demodulation at Microwave Frequencies. S. E. Harris, A. E. Siegman, and B. J. McMurtry. Polytechnic Institute of Brooklyn Symposia Series, XIII, Optical Masers, April 1963.

Several novel microwave light demodulators are described: traveling-wave microwave phototubes, frequency-modulation phototubes, and a birefringent descriminator which converts FM light to AM light. Microwave semiconductor photodiodes have also been tested. Microwave-modulated light, both AM and FM, can now be readily demodulated.

250. Investigation of Gas Lasers. T. S. Hartwick, AD289525, tp., September 1962.

A He-Ne gas laser was studied. An experimental investigation of a novel method for pumping the laser with a cold-cathode discharge was initiated.

251. Electronics Program. Nonlinear Optical Effects. T. S. Hartwick, AD289345, tp., October 1962.

The generation of subharmonics by an intense laser beam was studied theoretically and experimentally.

252. Limits on Directivity and Intensity of the Output of an Optical Maser. R. D. Haun, Jr., and R. C. Ohlmann, J. Opt. Soc. Am., Vol. 51, p. 473, April 1961.

An expansion of the treatment of Schawlow and Townes (Phys. Rev., Vol. 112, p. 1940) is given for the degree of directivity which can be obtained in the output of an optical maser. Expressions have been obtained which relate the maximum directivity of the output to the dimensions of the structure.

253. Noise in Optical Maser Amplifiers. H. A. Haus and J. A. Mullen, Polytechnic Institute of Brooklyn Symposia Series, XIII, Optical Masers, April 1963.

The definition of signal-to-noise ratio as necessitated by quantum considerations is discussed. The noise figure of an optical maser is derived. In quantum theory there is a distinction between the generators needed to represent thermal and zero-point mean square field (voltage) fluctuations and those representing noise power. At optical frequencies this distinction is of importance and forms the central theme of the paper.

254. Efficiency of Production of Intrinsic Recombination Radiation. J. R. Haynes, M. Lax, and W. F. Flood, Bull. Am. Phys. Soc., II, Vol. 6, p. 147, March 1961.

The efficiency is shown to be a function of the resistivity of the silica specimen, its temperature, carrier lifetime, and excess carrier density. Quantitative calculations of E are obtained using experimentally determined values of the radiative lifetime of excitons, binding energy, and radiative recombination rate of free electrons and holes.

255. Optical Masers. O. S. Heavens, Appl. Optics, Supplement 1, pp. 1-24, 1962.

Concepts basic to the understanding of optical masers are discussed, and a review of progress is made. Work on solid-state systems is summarized, and applications in the areas of fundamental experiments, communications, instrumentation, and surgery discussed.

256. Some Current Developments in Optical Masers. O. S. Heavens, J. Opt. Soc. Am., Vol. 53, p. 513, April 1963.

The coherence in the output of an optical maser is discussed. Various types of resonators are considered and the limits imposed on their performance through limitations in reflecting systems discussed. Possible methods of extending these limits are mentioned. Some results obtained with a new gas system are given.

257. Factors Affecting Population Inversion in Vibrational Levels by Overtone Pumping H. Heil, Bull. Am. Phys. Soc., II, Vol. 8, p. 443, 1963.

The unharmonicity of the bond of a diatomic molecule makes possible optical transitions involving $\Delta \nu = 2$ or more. If the fundamental is carefully removed from the pump radiation, one obtains a temporary population inversion.

258. Faraday Effect as a Q-Switch for Ruby Laser. J. L. Helfrich, J. Appl. Phys., Vol. 34, pp. 1000-1001, April 1963.

The successful application of the Faraday effect as a Q-switching technique for a ruby laser is reported. The single pulse obtained is 60 nsec wide and peak power is 600 KW.

259. Application of the Senarmont Polariscope to Analysis of Optical Maser Light. D. Hellerstein, J. Opt. Soc. Am.,Vol. 53, p. 515, April 1963.

The device employs an optically active wedge and an analyzing polarizer to map the azimuthal angle onto a linear field, indicating this angle by null positions on a photographic plate.

260. Theory of the Pulsation of Fluorescent Light from Ruby. R. W. Hellwarth, Phys. Rev. Lett., Vol. 6, pp. 9-12, January 1961.

The approach taken predicts that as long as the pump power is above a certain threshold, part of the fluorescent power will occur in recurrent bursts or pulses. From the theory are derived quantitative estimates of the pulse repetition rate, the fraction of power in the pulses, and the nature of the output between pulses, all in terms of the pump power and the ordinary properties of the crystal and plates.

261. Control of Fluorescent Pulsations. R. W. Hellwarth, pp. 334-341 in Advances in Quantum Electronics, J. R. Singer, ed., Columbia University Press, New York, 1961.

The coherent output of an optical maser may be caused to emit in intense, controllable pulses by varying the effective reflectivity of the cavity enclosure. Expected performance in the case of the ruby optical maser is evaluated.

262. Theory of Optical Harmonics. W. C. Henneberger, Bull. Am. Phys. Soc., II, Vol. 7, p. 14, January 1962.

The quantum theory of radiation is applied to the generation of optical harmonics by high-intensity beams of polarized monochromatic light. Expressions for the intensities of the first and second harmonics are given as functions of frequency and intensity of the incident light and atomic matrix elements.

263. Experimental Study of Gas Laser. G. Hepner, C.F.T.H., Paris, France, Third International Symposium on Quantum Electronics, Paris, France, February 1963.

The power of the coherent light of a 1.15 -micron neon-helium laser is studied in terms of the pressure of the gases and the electric parameters of the excitation source. The coherence of the beam is measured by interference and its polarization studied.

264. Relationship between the Near-Field Characteristics of a Ruby

Laser and Its Optical Quality. Michael Hercher, Appl. Optics, Vol. 1, pp. 665-670, September 1962.

An interferometric method of assessing crystal quality is described. The result of preliminary investigations of the spatial distribution of coherence in the near field of a good laser are given, and it is shown that there is a high degree of coherence between points separated by as much as 3.2 mm on the face of a ruby laser.

265. Optical Correction of Ruby Lasers. M. Hercher, J. Opt. Soc. Am., Vol. 52, p. 1319, 1962.

The laser rods were optically corrected. The improved quality of the emission pattern is evidenced by a more nearly uniform near-field energy distribution and by a well-defined ring structure with an intense central core in the far field (15 min).

266. Optical Properties of the Beam from a Continuous Helium-Neon Optical Maser. D. R. Herriott, J. Opt. Soc. Am., Vol. 51, p. 476, April 1961.

The beam from an optical maser operating in the near infra-red has been examined for its optical characteristics. Distribution of light in the near-field and far-field patterns of the beam have been determined. The shape and phase stability of the output wavefront is described.

267. Optical Properties of the Cavity and Output Beam of a Continuous Gas Maser. D. R. Herriott, pp. 44-49 in Advances in Quantum Electronics, J. R. Singer, ed., Columbia University Press, New York, 1961.

Optical characteristics of the cavity and the output beam of the gaseous helium-neon maser are considered.

268. Optical Properties of a Continuous Helium-Neon Optical Maser. D. R. Herriott, J. Opt. Soc. Am., Vol. 52, pp. 31-37, January 1962.

A continuous optical maser has been operated at five wavelengths in the near infrared. The optical amplification is provided by maser action in a discharge through a mixture of helium and neon gas. Examination of the beam shows that it is almost diffraction-limited for its one-centimeter diameter.

The spectral line shape at each transition is made up of three or more components, each less than a few hundred cycles in width, separated by the spacing of orders in the interferometer.

269. Coherent Generation of Light in Gas-Phase Chemical Reaction. M. Hertzberg, Polytechnic Institute of Brooklyn Symposia Series, XIII, Optical Masers, April 1963.

Coherent generation of light in gas-phase chemical reactions is discussed in terms of energy levels and transition probabilities. Using characteristic values for the reaction rates involved, it is known that chemiluminescent coherent generation is theoretically possible. Experimental studies of chemiluminescence excitation in alkali metal-halogen and alkali metal-group II halide reactions are in progress.

270. Detection of Ruby Laser Axial Mode Differences with Photodiodes. B. Herzog, A. P. Rodgers, J. E. Peterson, M. E. Lasser, G. Lucovsky, and R. B. Emmons, J. Opt. Soc. Am., Vol. 52, p. 594, May 1962.

Heterodyned difference signals between longitudinal laser optical cavity modes have been detected in photodiodes with high signal-to-noise ratio. Observations with the same photo-detection system have been made of the fundamental (approximately 1 kMc/sec) and its second and third harmonic from both $0°$ and $90°$ ruby rods.

271. The Concept of Coherence, Its Application to Lasers. H. Hodara, Hallicrafters, Chicago, Illinois, Third International Symposium on Quantum Electronics, Paris, France, February 1963.

The concept of coherence is reviewed critically and the operational definition in terms of visibility and persistence of interference is applied to lasers.

272. Broadband Masers Working at Room Temperature. T. A. Hoffman, S. Takacs, and T. Toth, Research Institute for Telecommunication, Budapest, Hungary, Third International Symposium on Quantum Electronics, Paris, France, February 1963.

Causes of maser noise in general, bandwidth and noise analysis, temperature dependence of the various types of noise, advantages of monatomic gaseous systems, and the

59

effect of optical pumping on noise and on bandwidth are discussed.

273. Masers and Lasers. C. A. Hogg and L. C. Sucsy, 226 pp., Maser/Laser Associates, Cambridge, 1962.

The history of the development of the laser is presented. A simplified review is made of the field of stimulated emission devices.

274. Coherent (Visible) Light Emission from $Ga(As_{1-x}P_x)$ Junctions. N. Holonyak, Jr., and S. P. Bevacqua, Appl. Phys. Lett., Vol. 1, pp. 82-84, December 1962.

Evidence for coherent light emission is based upon the observation of a threshold current beyond which the light intensity increases sharply, upon the pronounced narrowing of the spectral distribution of emitted light beyond threshold, and upon the sharply beamed radiation pattern of the emitted light. The results show that a p-n junction laser can be built with an output wavelength which can be selected in the range from 6200 A to near 8400 A.

275. Internally Reflecting Confocal Optical Maser Configuration. D. F. Holshouser, University of Illinois, Urbana, Illinois, Third International Symposium on Quantum Electronics, Paris, France, February 1963.

A configuration has been found which provides the advantages of internal reflection together with confocal properties. A description of this configuration and encouraging experimental results with a neodymium-in-glass optical maser are presented.

276. Microwave and Electro-Optical Properties of Liquids Exhibiting the Kerr Effect. D. F. Holshouser and R. Stanfield, University of Illinois, Urbana, Illinois, Third International Symposium on Quantum Electronics, Paris, France, February 1963.

The results of a program for the measurement of the electrical and optical properties of liquids used in Kerr cell light modulators are presented. Data and calculations are presented which permit comparison with solid-state electro-optical effects for light modulation at microwave frequencies.

277. Cooling to Very Low Temperatures by Means of Lasers. A. Honig, Bull. Am. Phys. Soc., II, Vol. 8, p. 233, March 1963.

A method to obtain temperatures down to about $0.05°$ K without employing a magnetic field is considered.

278. Laser-Induced Emission of Electrons, Ions and Neutral Atoms from Solid Surfaces. R. E. Honig and J. R. Woolston, Appl. Phys. Lett., Vol. 2, pp. 138-139, April 1963.

Studies of various solids irradiated in a vacuum with a focused beam from a pulsed ruby laser are described.

279. Maser Action in the Optical Range and the Possibility of Absolute Negative Absorption in Certain Molecular Spectra (the "Light Avalanche"). F. G. Houtermans, Helv. Phys. Acta, Vol. 33, pp. 933-940, 1960 (in German).

The conditions for three-level maser action are met by many molecular systems in which the upper state radiates to a negligibly populated intermediate level. Spectral distribution of the emission of the hydrogen continuum from H_2 is discussed and approximate expressions for the negative absorption coefficients are derived.

280. CW Operation of a GaAs Injection Laser. W. E. Howard, F. F. Fang, F. H. Dill, Jr., and M. I. Nathan, IBM Res. Dev., Vol. 7, pp. 74-75, January 1963.

Continuous stimulated emission of radiation from forward-biased GaAs diodes is reported.

281. Time Effects in GaAs Injection Lasers. W. E. Howard, F. F. Fang, F. H. Dill, Jr., and M. I. Nathan, Bull. Am. Phys. Soc., II, Vol. 8, p. 88, 1963.

The spectral distribution of stimulated emission from GaAs junctions is observed to vary strongly with time. This has been studied by varying both the length and the frequency of the input current pulses. Diodes were sometimes immersed in liquid He II. For a given pulse, the general result is that the emissions shift to longer wavelengths; peaks corresponding to different modes do not themselves shift, but rather their relative intensities change to favor longer-wavelength modes.

282. Optical-Pumping Cavity Construction Technique. R. H. Hronik, R. C. Jones, and C. J. Bronco, Rev. Sci. Instr. Vol. 33, pp. 766-777, July 1962.

A technique for constructing lightweight elliptical reflectors of any size by the use of simple settings is described.

283. Parametric Photon Interactions and Their Applications. H. Hsu and K. F. Tittel, General Electric Company, Syracuse, N. Y., Lasers and Applications Symposium, Ohio State University, November 1962.

> The recent achievement of obtaining second and third harmonic radiation at optical frequencies with an intense monochromatic laser beam has led to great interest in phenomena resulting from nonlinear interactions between radiation and matter. The mechanisms of these harmonic generations can be identified as parametric interactions. In particular, the concept of traveling-wave parametric interaction can be applied to enhance these interactions, as was demonstrated by Giordmaine, and Maker et al. Furthermore, these interactions can be interpreted as typical examples of three-dimensional parametric interactions of photons as quasi-particles.

284. Photoconductive Mixing Experiments Using a Ruby Laser and a Semiconductor Microwave Photodiode. H. Inaba and A. E. Siegman, Stanford Electronics Laboratories, Tech. Rept. 177-2, June 1962.

> This report describes the operation of semiconductor microwave photodiodes as an optical frequency mixer to produce microwave photo beats from the light output of a ruby optical maser.

285. Microwave Photomixing of Optical Maser Outputs with a PIN-Junction Photodiode. H. Inaba and A. E. Siegman, Proc. IRE, Vol. 50, p. 1823, August 1962.

> This paper reports the possibility of using a PIN-junction photodiode as a mixer in an optical superheterodyne system employing microwave-modulated light.

286. The Effect of Strong Illumination of the Absorptivity of Complex Molecules. A. P. Ivanov, Optics and Spectroscopy, Vol. 8, pp. 183-188, March 1960.

> Formulas are derived for the absorptivity of systems with two and three energy levels. Conditions under which negative ab-

sorption takes place are examined. It is shown that it is possible to observe clarification of a substance.

287. The Spherical Fabry-Perot Interferometer as an Instrument of High Resolving Power for Use with External or with Internal Atomic Beams. D. A. Jackson, Proc. Roy. Soc., A, Vol. 263, pp. 289-308, 1961.

> The spherical Fabry-Perot interferometer was designed by P. Connes as an instrument capable of realizing higher resolving power by virtue of its greater light power at high resolution and the much lower requirement with regard to accuracy of adjustment.

288. Coherent Light Amplification in Optically Pumped Cs Vapor. S. Jacobs, G. Gould, and P. Rabinowitz, Phys. Rev. Lett., Vol. 7, pp. 415-418, December 1961.

> Spatially coherent amplification in cesium vapor excited by selective optical pumping is measured.

289. Criteria for Optical Maser Amplifiers and Oscillators. H. Jacobs, D. Holmes, L. Hatkin, and F. A. Brand, U. S. Army Electronics Research and Development Laboratory, Fort Monmouth, N. J., Third International Symposium on Quantum Electronics, Paris, France, February 1963.

> The ruby optical maser amplifier is treated as an electromagnetic boundary value problem. An equivalent transmission line analogue has been formulated. A portion of this line, the ruby crystal has negative attenuation. Criteria for maximum gain and oscillation have been established. A backward wave amplifier with higher stability but lower gain is possible.

290. Optical Heterodyning with a CW Gaseous Laser. S. Jacobs and P. Rabinowitz, TRG, Syosset, New York, Third International Symposium on Quantum Electronics, Paris, France, February 1963.

> A modified Twyman-Green interferometer is used to divide the laser beam into signal and local oscillator beams, which are recombined and mixed on the surface of a square law photodetector. Information and attenuation are impressed on the signal beam prior to detection. Measurements of signal noise were made as a function of spatial coherence between local oscillator and signal beam, local oscillator beam power,

detector quantum efficiency, and multimoding.

291. Optical Heterodyne Detection of Coherent Light. S. F. Jacobs, P. Rabinowitz, J. T. LaTourrette, and G. Gould, J. Opt. Soc. Am., Vol. 53, p. 515, April 1963.

> Experimental results show that signal-to-noise ratio is preserved provided the local oscillator current is greater than the dark current and that under these conditions the signal-to-noise power ratio corresponds to the theoretical value.

292. Stability and Resettability of He-Ne Masers. T. S. Jaseja, A. Javan, J. Murray, and C. H. Townes, Bull. Am. Phys. Soc., II, Vol. 7, p. 553, November 1962.

> The frequency stability of two He-Ne masers under quiet acoustical and thermal conditions has been examined. A beat frequency with short-term frequency width less than 100 cps was obtained over a period of 4 seconds.

293. Helium-Neon Optical Maser: A New Tool for Precision Measurements. T. S. Jaseja and A. Javan, Massachusetts Institute of Technology, Cambridge, Mass., Lasers and Applications Symposium, Ohio State University, November 1962.

> The frequency characteristics, such as stability and resettability, of a cw He-Ne optical maser have been studied in some detail. An rf beat with a total short-term frequency width less than 100 cps was obtained between the two masers. This represents a frequency stability better than 3 parts in 10^{13} during a time interval of a few seconds. Under quiet acoustical and thermal conditions, the average frequency of this beat due to thermal fluctuations varied less than 50 cps during a one-second period. This indicates that the optical maser is capable of detecting changes in length smaller than 2 parts in 10^{13}.

294. Frequency Stability of He-Ne Masers and Measurements of Length. T. S. Jaseja, A. Javan, and C. H. Townes, Phys. Rev. Lett., Vol. 10, pp. 165-167, March 1963.

> Frequency stability of the oscillations of He-Ne masers at 1.153 microns has been examined under more controlled conditions than was previously the case. Frequency spread of the oscillation has been reduced to about 20 cps, or about

8 parts in 10^{14}.

295. Possibility of Production of Negative Temperature in a Gas Discharge. A. Javan, Phys. Rev. Lett., Vol. 3, pp. 87-89, June 1959.

   An analysis of possible systems capable of producing negative temperatures is presented.

296. Possibility of Obtaining Negative Temperature in Atoms by Electron Impact. A. Javan, pp. 564-571 in Quantum Electronics, C. H. Townes, ed., New York, Columbia University Press, 1960.

   Theoretical possibilities for negative temperature between two atomic levels in a gas are discussed. An experiment for the detection of this effect is described.

297. Optical Maser Oscillations in a Gaseous Discharge. A. Javan, pp. 13-27 in Advances in Quantum Electronics, J. R. Singer, ed., Columbia University Press, New York, 1961.

   The processes involved in the discharge leading to the inverted populations in the Ne levels are described.

298. Frequency Characteristics of a Continuous Wave He-Ne Optical Maser. A. Javan, E. A. Ballik, and W. L. Bond, J. Opt. Soc. Am., Vol. 52, pp. 96-98, January 1962.

   The results of experimental studies concerning the spectral purity and frequency stability of the light output of two cw He-Ne optical masers are given.

299. Population Inversion and Continuous Optical Maser Oscillation in a Gas Discharge Containing a He-Ne Mixture. A. Javan, W. R. Bennett, Jr., and D. R. Herriott, Phys. Rev. Lett., Vol. 6, pp. 106-110, February 1961.

   The results of experimental determinations of several physical properties of a gaseous discharge consisting of a He-Ne mixture which have led to the successful operation of a continuous wave maser at five different wavelengths in the near infrared.

300. Short-Time Frequency Stability of He-Ne Optical Masers. A. Javan, T. S. Jaseja, and C. H. Townes, Bull. Am. Phys. Soc., II, Vol. 8, pp. 380-381, April 1963.

The spectra of two independent masers were examined in a carefully controlled and acoustically isolated environment. Their beat frequency was heterodyned to an audio frequency and recorded on magnetic tape.

301. Optical Maser Characteristics of Nd^{3+} in CaF_2. L. F. Johnson, J. Appl. Phys. (Correspondence), Vol. 33, p. 756, February 1962.

  Optical maser   effects are observed at 1.046 microns.

302. Continuous Operation of the $CaF_2:Dy^{2+}$ Optical Maser. L. F. Johnson, Proc. IRE, Vol. 50, pp. 1691-1692, July 1962.

  This letter reports lifetime and spectroscopic data on divalent dysprosium in CaF_2 and continuous operation of the optical maser at 20°K.

303. Optical Maser Characteristics of Rare Earth Ions in Crystals. L. F. Johnson, J. Appl. Phys., Vol. 34, pp. 897-909, April 1963.

  Several divalent and trivalent rare earth ions incorporated in various host crystals have been found to exhibit stimulated emission in the near infrared. A report is presented describing some of the basic characteristics of these materials; absorption and fluorescence spectra, energy level diagrams, optical maser wavelengths, and operating temperatures and thresholds for stimulated emission.

304. Optical Maser Characteristics of Trivalent Thulium in Calcium Tungstate. L. F. Johnson, G. D. Boyd, and K. Nassau, Proc. IRE, Vol. 50, pp. 86-87, January 1962.

  Optical stimulated emission in the fluorescence line at 1.911 microns is observed at 77°K. An energy level diagram depicting the optical maser action is proposed.

305. Optical Maser Characteristics of Ho^{3+} in $CaWO_4$. L. F. Johnson, G. D. Boyd, and K. Nassau, Proc. IRE, Vol. 50, pp. 87-88, January 1962.

  Stimulated emission in the infrared fluorescence line at 2.046 microns is observed at 77°K. An energy level diagram depicting the optical maser action is proposed.

306. Continuous Operation of the $CaWO_4$:Trivalent Neodymium Optical

Maser at Room Temperature. L. F. Johnson, G. D. Boyd, and K. Nassau, J. Opt. Soc. Am., Vol. 52, p. 608, May 1962.

Laser operation is reported at room temperature with flowing water as the crystal coolant. Threshold power is approximately twice that required at liquid oxygen temperature.

307. Continuous Operation of the $CaWO_4:Nd^{3+}$ Optical Maser. L. F. Johnson, G. D. Boyd, K. Nassau, and R. R. Soden, Proc. IRE, Vol. 50, p. 213, February 1962.

Continuous operation of a solid-state optical maser has been demonstrated. The system investigated is $CaWO_4$ containing trivalent neodymium. Generation of a continuous coherent electromagnetic wave at 1.065 microns is obtained.

308. Continuous Operation of a Solid-State Optical Maser. L. F. Johnson, G. D. Boyd, K. Nassau, and R. R. Soden, Phys. Rev., Vol. 126, pp. 1406-1409, May 1962.

Experiments demonstrating continuous operation of a solid-state optical maser are described. A continuous coherent wave at 1.065 microns is obtained from trivalent neodymium in $CaWO_4$ operating at about 85° K.

309. Piezoelectric Optical Maser Modulator. L. F. Johnson and D. Kahng, J. Appl. Phys., Vol. 33, pp. 3440-3443, December 1962.

An optical maser modulator employing a transparent piezoelectric medium is described. The device is based on multiple interference in transmission or reflection, the intensity in the interference pattern being modulated by the change in thickness of the piezoelectric plate. Using barium titanate it is shown that 80% modulation and 20 mc bandwidths are readily attainable.

310. Optical Maser Characteristics of Nd^{3+} in $SrMoO_4$. L. F. Johnson and R. R. Soden, J. Appl. Phys. (Correspondence), Vol. 33, p. 757, February 1961.

Stimulated emission at 1.064 microns is observed.

311. Optical Maser Characteristics of Trivalent Neodymium in Calcium Fluoride and Strontium Molybdate. L. F. Johnson and R. R. Soden, Bull. Am. Phys. Soc., II, Vol. 7, p. 14, January 1962.

Optical maser action is reported at 1.0643 microns at room temperature and at 1.064 microns at 77°K in SrMgO$_4$ and at 77°K in CaF$_2$.

312. A Ruby Laser Exhibiting Periodic Relaxation Oscillations. R. E. Johnson, W. H. McMahan, F. J. Oharek, and A. P. Sheppard, Proc. IRE, Vol. 49, pp. 1942-1943, December 1961.

A ruby optical maser with confocal ends and optical axis oriented 90° to the cylinder axis has been constructed. The periodic characteristics of the output pulse should be useful in optical ranging applications using correlation techniques.

313. A Laser Satellite Tracking Experiment. T. S. Johnson and H. H. Plotkin, NASA Goddard Space Flight Center, Lasers and Applications Symposium, Ohio State University, November 1962.

An experiment is described which utilizes a ruby optical maser to illuminate a satellite for tracking purposes. The components of the tracking system are described and the performance of the system is analyzed. The transmitter is a ruby optical maser employing Q-switching to obtain a short-duration, high-peak-power pulse. The receiver is a high-gain photomultiplier with enhanced sensitivity at the ruby output wavelength.

314. Optical Maser Oscillation in Ruby. K. Kabota and K. Hayashi, J. Phys. Soc. Japan, Vol. 16, p. 2063, 1961.

Linde Al$_2$O$_3$:0.05% Cr$_2$O$_3$ was excited by a spiral xenon flash and emission of a 6943 A light was observed with relaxation oscillation when the input energy exceeded a certain threshold value.

315. Fluorescence and Optical Maser Effects in CaF$_2$:Sm^{++}. W. Kaiser, C. G. Garrett, and D. L. Wood, Phys. Rev., Vol. 123, pp. 766-776, August 1961.

Measurements are reported of absorption, emission, and activation spectra in CaF$_2$:Sm^{2+} and also of the fluorescence lifetime. A revised level scheme is proposed.

316. Two Photon Excitation in CaF$_2$:Eu^{++}. W. Kaiser and C. G. Garrett, Phys. Rev. Lett., Vol. 7, pp. 229-231, September 1961.

The generation of blue fluorescent light around 4250 A by illuminating $CaF_2:Eu^{2+}$ crystals with red light of 6943 A is investigated.

317. Fluorescence and Optical Maser Effects in CaF_2:Sm. W. Kaiser, C. G. Garrett, and D. L. Wood, Phys. Rev., Vol. 123, pp. 766-776, August 1961.

Measurements are reported of absorption, emission, and activation spectra, and fluorescence lifetime. A revised level scheme is proposed. Observations of optical maser effects were made at liquid helium and liquid nitrogen temperatures over a wide range of pumping intensities. For illumination in the 6400 A band, the threshold intensity of illumination was about 20 W/m^2. Observations are reported of the dependence of the intensity of the maser beam on the pumping intensity. Five distinct frequencies were observed. The number of modes excited was of the order of 1000.

318. Scattering Losses in Optical Maser Crystals. W. Kaiser and M. J. Keck, J. Appl. Phys. (Correspondence), Vol. 33, pp. 762-764, February 1962.

Attention is drawn to one specific crystal imperfection, namely, small foreign particles within the crystal. It will be shown that the extinction losses resulting from optical scattering on these small particles can be experimentally determined and they can be estimated from the number and size of scattering centers.

319. Splitting of the Emission Lines of Ruby by an External Electric Field. W. Kaiser, S. Sugano, and D. L. Wood, Phys. Rev. Lett., Vol. 6, pp. 605-607, June 1961.

Upon superimposing an external electric field upon the internal crystal field a new additional splitting of the emission lines is observed which is much larger than expected from a superficial consideration of the Stark effect.

320. Stimulated Light Emission. H. Kallman, Polytechnic Institute of Brooklyn Symposia Series, XIII, Optical Masers, April 1963.

Stimulated light emission is discussed in terms of classical electromagnetic theory as well as in terms of quantum theory. Its importance for Planck's radiation law and for dispersion theory is shown. Stimulated emission escaped direct detection from the classical viewpoint, since in a system of harmonic oscillators stimulated emission does not manifest itself directly. The report describes how its extension and its coherence were eventually proved in direct observation by measuring the anomalous dispersion of excited states.

321. Proposed Technique for Modulation of Coherent Light. A. K. Kamal and S. D. Sims, Proc. IRE, Vol. 49, p. 1331, August 1961.

A technique for frequency modulation of the coherent light output of a ruby laser utilizing the Stark effect of the ruby itself is proposed.

322. Effect of Temperature on Laser Output. A. K. Kamal and S. D. Sims, Presented at Fourth International Conference on Microwave Tubes, The Hague, Netherlands, 1962.

An investigation of the effects of temperature on the output of a pulsed ruby laser has been performed. The starting and ending time, relative to the application of the pump, of the maser action was investigated and the pulse length was obtained.

323. Laser Power and Energy Measurement Using Nonlinear Polarization in Crystals. A. K. Kamal and M. Subramanian, Polytechnic Institute of Brooklyn Symposia Series, XIII, Optical Masers, April 1963.

When a laser beam propagates through a crystalline medium that lacks inversion symmetry, the very-high-intensity electric field of the radiation develops a dc polarization in the medium which is directly proportional to the instantaneous power in the laser beam. This principle has been used in demonstrating that a device can be built to measure the power in a laser pulse. The device is of the transmission type, and it enables one to measure the peak power in the pulse. Energy can be determined by passing the output through an integrator.

324. Temperature Dependence of the Complex Dielectric Constant in KH_2PO_4-Type Crystals and the Efficiency of Microwave Light Modulators. I. P. Kaminow, Bell Telephone Laboratories, Holmdel, New Jersey, Third International Symposium on Quantum Electronics, Paris, France, February 1963.

Measurements of the temperature dependence of the complex dielectric constants of KH_2PO_4 and several isomorphs at 9.2 Gc are reported.

325. Electro-Optical Light Modulation. I. Kaminow, Bell Telephone Laboratories, Lasers and Applications Symposium, Ohio State University, November 1962.

Some aspects of the design of microwave light modulators based on the electro-optic effect are reviewed. Recent work on cavity-type and traveling-wave-type microwave structures as well as measurements on electro-optic materials are presented.

326. The Power Chart for Evaluation of Modes in a Laser Oscillator. R. A. Kaplan, Polytechnic Institute of Brooklyn Symposia Series, XIII, Optical Masers, April 1963.

A chart which describes the model characteristics of a resonator and the gain of the laser materials has been developed. This chart, termed the "power chart" provides a graphical method of determining the modes of oscillation of a given laser as a function of pump power. It therefore permits rapid determination of the radiation beamwidth, bandwidth, and spectrum of the laser oscillator. It may also be employed to evaluate the effects of variations in the parameters of the laser on the performance of that device.

327. Model for Transient Oscillations in a Three-Level Optical Maser. J. I. Kaplan and Robert Zier, J. Appl. Phys.,Vol. 33, pp. 2372-2375, July 1962.

The transient response of a three-level optical maser is calculated for a density of homogeneously broadened atoms coupled to one homogeneously broadened cavity mode. The atoms are treated in terms of the density matrix and the cavity mode is treated classically. Numerical solutions are found for a certain set of parameters. The response line width is found assuming one cavity mode coupled to a density of inhomogeneously broadened atoms.

328. Soviet Laser Research. Kassel, Simon; Proc. IEEE, Vol. 51, pp. 216-218, 1963.

The author discusses Soviet laser research progress. Starting from early research, which terms masers "molecular amplifiers," later and current programs are discussed. Paramagnetic materials and infrared gas masers are described.

329. Atomes à l'Intérieur d'un Interféromètre Perot-Fabry. A. Kastler, Appl. Optics, Vol. 1, pp. 17-24, January 1962.

In the case of external illumination, the distribution of light intensity inside a Perot-Fabry interferometer is calculated. Local high intensity in the stationary waves inside can be much higher than that of the incident light beam. The realization of a laminar fluorescent medium is discussed. It is shown that this device is equivalent to a long absorption path in an ordinary light beam.

330. Combination of Optical Pumping and Magnetic Resonance Techniques. Application to Ions in Crystals. Alfred Kastler, Laboratoire de Physique, E.N.S., Paris, France, Lasers and Applications Symposium, Ohio State University, November 1962.

By combining optical pumping and magnetic resonance techniques interesting results have been obtained with monatomic vapors; the double resonance method introduced by Brossel led to the study of excited atomic states. The method has been extended to ground states and has received many applications: atomic orientation, nuclear orientation, studies of atomic and nuclear resonance and relaxation, collisions, and so on.

331. Optical Masers Utilizing Multiple Ruby Sections in Spherical Resonator. M. Katzman and O. Leifson, J. Opt. Soc. Am., Vol. 52, p. 602, May 1962.

It was determined that the radiation internal to the resonant structure consists of two cones focused at the common center of curvature of the two reflectors. The immediate result was that the pulse duration increased by a factor of two.

332. Proposal for Modulating the Output of an Optical Maser. P. Kaya, Proc. IRE, Vol. 50, p. 323, March 1962.

The use of the strong magnetic field of an intense coherent radiant energy beam to modulate a coherent plane-polarized beam of radiant energy at higher frequency and in particular the utilization of the Faraday effect are discussed.

333.   A Solution for Continuous Pumping of Solid-State Lasers. P. H. Keck, Bull. Am. Phys. Soc., II, Vol. 7, p. 15, January 1962.

>   It is shown that a continuous pumping flux which is above the threshold value required for pumping solid-state laser materials such as ruby can be obtained using a xenon arc lamp or the sun as a light source.

334.   Pumping Requirements for CW Solid-State Lasers. P. H. Keck, Bull. Am. Phys. Soc., II, Vol. 7, p. 118, February 1962.

>   The conditions for continuous pumping of a solid-state laser material depending on the spectral distribution of the source, absorption characteristics, threshold for laser action, and geometry are discussed. Several suggested arrangements for cw solid-state lasers are described.

335.   Continuous Solid-State Laser. P. H. Keck, J. Opt. Soc. Am., Vol. 52, p. 602, May 1962.

>   A light source is described which is capable of delivering continuously into a small laser crystal a luminous flux greater than the required threshold of most known solid-state laser materials. The source consists of a high-pressure Xe arc lamp, and a powerful optical system for concentrating a large flux into a small cross-sectional area.

336.   Comparison of Flash and CW Operation of Neodymium-Doped Calcium Tungstate Lasers. P. H. Keck, J. J. Redman, and C. E. White, Texas Instruments, Dallas, Texas, Third International Symposium on Quantum Electronics, Paris, France, February 1963.

>   A laser pumping system has been developed which permits both flash and cw pumping without changes of the light source, optical system, or laser rod geometry. A quantitative evaluation of the efficiency of different condenser systems has been obtained.

337.   Performance of a CW Neodymium Laser. P. H. Keck, J. J. Redman, C. E. White, and D. E. Bowen, J. Opt. Soc. Am., Vol. 52, p. 1323, 1962.

The high-pressure compact xenon arc lamp with an ellipsoidal collector of 24-in. diameter and an opening of f/0.4
served as a pumping source. The laser rod was cut from a
neodymium-doped calcium-tungstate single crystal and had
multiple dielectric coatings on both ends. The performance
characteristics of the laser under various conditions are
discussed.

338. New Condenser for a Sun-Powered Continuous Laser. P. H. Keck,
J. J. Redman, C. E. White, and R. E. Dekinder, J. Opt. Soc. Am.,
Vol. 52, p. 1319, 1962.

The present design yields a very uniform and most effective
excitation. The condenser consists of a conical part made
of a high-refractive-index material with a field lens on the
surface pointing toward the parabolic collector. The repeated reflections from the sphere walls cause the laser rod
to be pumped very uniformly.

339. Doping of Semiconductors for Injection Lasers. R. J. Keyes, Proc.
IEEE, Vol. 51, p. 602, April 1963.

The communication points out that the thermodynamic condition for stimulated emission in a semiconductor places a
restriction on the doping levels required for a lasing junction.

340. Recombination Radiation Emitted by Gallium Arsenide. R. J. Keyes
and T. M. Quist, Proc. IRE, Vol. 50, pp. 1822-1823, August 1962.

When appropriately diffused GaAs diodes are biased in the
forward direction they emit intense line radiation corresponding to gas transitions. Absolute measurements of the emitted radiation intensity indicate that at 77° K these diodes may
be as high as 85% efficient in the conversion of injected holes
into photons of the gap energy. Data pertaining to the spectral distribution and speed of response of the emitted radiation are presented, and the high conversion efficiency of the
diode and its implications are discussed.

341. Recombination Radiation Emitted by Gallium Arsenide. R. J.
Keyes and T. M. Quist, Lincoln Laboratory MIT, Lexington, Mass.,
Third International Symposium on Quantum Electronics, Paris,
France, February 1963.

When diffused GaAs diodes are biased in the forward direction at 77°K nearly one photon is emitted for each injected carrier. The emitted radiation is primarily concentrated in a narrow spectal band centered at 1.45 electron volts. Spectral distribution of radiation is a function of temperature and injection current. Radiative recombination time is less than 5×10^{-9} seconds.

342. Spin-Lattice Relaxation in the $2P_3(2E)$ State of Ruby. A. Kiel, pp. 417-425 in Advances in Quantum Electronics, J. R. Singer, ed., Columbia University Press, New York, 1961.

The theory of spin-lattice relaxation of the paramagnetic ground state ions is extended to excited states. A general method is applied to relaxations with the $2P_3$ manifold in ruby.

343. Multi-Phonon Spontaneous Emission in Paramagnetic Crystals. A. Kiel, Carlyle Barton Laboratory, The Johns Hopkins University, Third International Symposium on Quantum Electronics, Paris, France, February 1963.

As a possible explanation of the nonradiative transitions occurring between ionic states of paramagnetic crystals, the magnitude of the probability of multi-phonon emission processes in the crystals is considered. Specifically, higher-order perturbation theory is applied to determine the way the transition probability falls off for higher-order (multi-phonon) spontaneous emission of optical or acoustical phonons.

344. Alignment Characteristics of a Helium-Neon Optical Maser. J. Killpatrick, H. Gustafson, and L. Wold, Proc. IRE, Vol. 50, p. 1521, June 1962.

Laser output is investigated as a function of mirror alignment.

345. Stimulated Emission from p-n Junctions. J. D. Kingsley, General Electric Research Laboratory, Schenectady, New York, Third International Symposium on Quantum Electronics, Paris, France, February 1963.

Stimulated emission from forward-biased p-n junctions has been observed in the near infrared. This is manifested by a spectral narrowing from an emission band with a width

75

of 150 A to several lines each less than 0.5 A in width. Simultaneously the radiation pattern narrows in certain directions, implying that different parts of the p-n junction are radiating coherently with each other.

346. Spectral Characteristics of Stimulated Emission from GaAs Junctions. J. D. Kingsley, G. E. Fenner, and R. N. Hall, General Electric Research Laboratory, Lasers and Applications Symposium, Ohio State University, November 1962.

> Stimulated emission from forward-biased GaAs p-n junctions has been observed at approximately 8400 A, the specific wavelength depending on the construction of the diode. That the diode is emitting coherently is manifested by dramatic changes in the radiation pattern by a large decrease in spectral width as the diode current is increased. Under the best conditions obtained thus far a large fraction of the radiation is observed to be contained in a beam only 1° wide and 4° high.

347. Parametric Amplification and Oscillation at Optical Frequencies. R. H. Kingston, Proc. IRE, Vol. 50, p. 472, April 1962.

> It is proposed that coherent optical energy may be generated at subfrequencies if a nonlinear dielectric material is driven by an optical maser "pump" such as a ruby. The conditions for oscillation in a simple resonant system based on the observed experimental data for second harmonic generation are derived.

348. A Pulsed Ruby Maser as a Light Amplifier. P. P. Kisliuk and W. S. Boyle, Proc. IRE, Vol. 49, pp. 1634-1639, November 1961.

> Power amplification of a factor of two has been achieved for visible light. Observation of amplification has been obtained in the helium-neon gas device but the gain is considerably less than two. The dependence of the gain on temperature and pumping power agrees with theory within experimental limits.

349. Energy Levels in Concentrated Ruby. P. Kisliuk, A. L. Schawlow, and M. D. Sturge, Bell Telephone Laboratories, Murray Hill, N.J., Third International Symposium on Quantum Electronics, Paris, France, February 1963.

> The temperature dependence of absorption in the N_1 and N_2

lines of concentrated ruby has been studied. A tentative assignment has been made for the sign and magnitude of the exchange coupling and for the total spin in the lower state of the transition. Other observed lines can be assigned to transitions between the same upper level and lower levels of different total spin, in good agreement with the interval rule and consistent with their temperature dependence.

350. The Interference Between Beams from the Opposite Ends of a Ruby Optical Maser. P. Kisliuk and D. J. Walsh, Appl. Optics, Vol. 1, pp. 45-49, January 1962.

Interference fringes have been observed when a pulsed ruby optical maser has been constructed which allows the beams from opposite ends to be superimposed. Interference fringes have been observed which can be interpreted to show that the coherence predicted by theory is in fact observed.

351. "Hole-Burning" Effect on the Output Frequency of a Ruby Optical Maser. P. Kisliuk and D. J. Walsh, Bull. Am. Phys. Soc., II, Vol. 7, p. 330, April 1962.

For input energy considerable in excess of threshold under conditions for which the number of modes contained within the line width is very great, the output of a ruby optical maser consists of several bands of frequencies. The number and spacing of the bands are dependent on pumping energy and temperature. This is explained as a result of hole burning in the excited state.

352. Energy Levels of Divalent Dysprosium in CaF_2, BaF_2, and SrF_2. Z. J. Kiss, Bull. Am. Phys. Soc., II, Vol. 7, p. 445, August 1962.

From the fluorescence and absorption spectra the energy levels of divalent dysprosium were determined in the three crystal lattices. The cubic crystal field splitting of the 518 and 517 states was observed, and the fourth and sixth order terms of the cubic field potential were deduced. The symmetry of the states was confirmed by Zeeman-effect studies of the fluorescent lines. The 2.36-micron laser transition in the CaF_2 system was identified and was correlated with other spectroscopic data.

353. Energy Levels of Divalent Thulium in CaF_2. Z. J. Kiss, Phys. Rev., Vol. 127, pp. 718-724, August 1962.

From the fluorescence and absorption spectra and energy level scheme for the system is proposed.

354. Zeeman Tuning of the $CaF_2:Tm^{2+}$ Optical Maser. Z. J. Kiss, Appl. Phys. Lett., Vol. 2, pp. 61–62, February 1963.

   The tuning of an optical maser with a magnetic field over a range of 1.5 cm^{-1} is demonstrated.

355. Zeeman Effect, Stark Effect, and Line Width of the Optical Maser Transition in $CaF_2:Tm^{2+}$. Z. J. Kiss, RCA Laboratories, Princeton, N. J., Third International Symposium on Quantum Electronics, Paris, France, February 1963.

   The optical maser operates in two linearly polarized components of the four-line Zeeman pattern which can be tuned with an effective value of g equal to two with respect to each other. A second-order Stark shift of the above transition was observed. A study of the fluorescent line width, its temperature dependence, and its origin is discussed.

356. Divalent Rare-Earth in CaF_2 as Optical Maser Materials. Z. J. Kiss, RCA Laboratories, Princeton, N. J., Lasers and Applications Symposium, Ohio State University, November 1962.

   From the absorption and fluorescence spectra of divalent rare-earth-doped CaF_2 a tentative energy level diagram for the 4f and 5d-6s, etc., levels is deduced, and the possibilities for maser action in the various systems are discussed. The characteristics of the 4f-4f magnetic dipole and the 5d-4f electric dipole transitions are compared. The behavior of the divalent rare earths in calcium fluoride optical maser systems is presented.

357. Optical Maser Action in $CaWO_4:Er^{3+}$. Z. J. Kiss and R. C. Duncan, Jr., Proc. IRE, Vol. 50, p. 1531, June 1962.

   Maser action is reported at 1.612 microns. An energy level diagram showing the maser transition is included.

358. Pulsed and Continuous Optical Maser Action in $CaF_2:Dy^{2+}$. Z. J. Kiss and R. C. Duncan, Jr., Proc. IRE, Vol. 50, pp. 1531–1532, June 1962.

   Maser action is reported at 2.36 microns. An energy level

diagram showing the maser transition is included.

359. Optical Maser Action in $CaF_2:Tm^{2+}$. Z. J. Kiss and R. C. Duncan, Jr., Proc. IRE, Vol. 50, pp. 1532-1533, June 1962.

> Optical maser action is reported at 1.116 microns at liquid helium temperature.

360. Sun-Pumped Continuous Optical Maser. Z. J. Kiss, H. R. Lewis, and R. C. Duncan, Jr., J. Appl. Phys. Lett., Vol. 2, pp. 93-95, March 1963.

> Optical maser action in the $CaF_2:Dy$ system at liquid neon temperature using the sun as a pumping source is reported.

361. Ring Laser Device Performs Rate Gyro Angular Sensor Functions. P. J. Klass, Avionics, Vol. 78, pp. 98-99, February 1963.

> Laser gyro uses four gas lasers in ring configuration to measure angular rate from the Doppler shift produced on contrarotating beams.

362. Laser and Two-Photon Processes. D. A. Kleinman, Phys. Rev., Vol. 125, pp. 87-88, January 1962.

> The possibility of observing two-photon processes using the intense beam of a ruby optical maser is considered theoretically for two general types of experiment. The essential approximation in the theory is the hypothesis that excited states exist which connect both the initial and final states by electronic dipole transitions having a total oscillator strength f-1. Simple formulas are obtained for two-photon absorption which do not depend on the details of electronic structure. Recent experiments in the $CaF_2:Eu^2$ system are in quantitative agreement with the predictions of the theory.

363. Discrimination Against Unwanted Orders in the Fabry-Perot Resonator. D. A. Kleinman and P. P. Kisliuk, B.S.T.J., Vol. 41, pp. 453-462, March 1962.

> It is proposed that the usual Fabry-Perot interferometer structure of the optical maser may be modified in a very simple way to provide discrimination against unwanted orders. The modification is an extra reflecting surface suitably positioned outside the maser which can greatly affect the losses

of the various orders. A simple one-dimensional analysis is given for the effect, and numerical results are presented for a realistic case, showing that the effect can be large.

364. The Thermal Resistivity of Ruby in the Optically Excited State. P. G. Klemens, Appl. Phys. Lett., Vol. 2, pp. 81-83, February 1963.

Estimates are made of the increase in thermal resistance due to transitions between the two levels of the R doublet and the resultant emission or absorption of a phonon.

365. Properties of Hydrogen Maser. D. Kleppner, H. M. Goldenberg, and N. F. Ramsey, Appl. Optics, Vol. 1, pp. 55-60, January 1962.

Properties of the hydrogen maser and details of the apparatus are discussed. Analysis of a maser oscillator is given in detail.

366. Transmission of Ruby Laser Light Through Water. G. L. Knestrick and J. A. Curcio, J. Opt. Soc. Am., Vol. 53, p. 514, April 1963.

Attenuation was measured over a distance of 6 to 63 meters at wavelengths of 6943 and 4900 A. The attenuation coefficients were 5.5×10^{-3} and 8.6×10^{-4} cm^{-1}, respectively.

367. Three-Reflector Optical Cavity for Mode Discrimination. T. R. Koehler and J. P. Goldsborough, Bull. Am. Phys. Soc., II, Vol. 7, p. 446, August 1962.

The use of a Fabry-Perot resonator as an end reflector is analyzed and optimum operating conditions are obtained. Beam narrowing and a reduction in mode density have been observed experimentally.

368. Optimizing the Parameters for an End-Pumped Laser. C. J. Koester and D. A. LaMarre, J. Opt. Soc. Am., Vol. 52, p. 595, May 1962.

The conditions for continuous laser action have been derived for the case of a multiplicity of upper and lower laser levels and an index of refraction other than unity. Parameters which can be optimized are the length, diameter, doping, and side wall reflecting material.

369. Experimental Laser Retina Coagulator. C. J. Koester and E. Snitzer

J. Opt. Soc. Am., Vol. 52, p. 607, May 1962.

> A ruby laser has been used as the light source for a photo-coagulation of the human retina. Laser properties which are pertinent to retinal coagulation are the energy output, transmittance of the ocular media, and absorption by the fundus at the laser wavelength, and the beam diameter and spread.

370. Fiber Laser as a Light Amplifier. C. J. Koester and E. Snitzer, J. Opt. Soc. Am., Vol. 53, p. 515, April 1963.

> Fiber lasers in helix form and straight section have been used as amplifiers of 1.06 micron radiation.

371. Interactions Between Two Nd^{3+} Glass Lasers. C. J. Koester, R. F. Woodcock, Elias Snitzer, and H. M. Teager, J. Opt. Soc. Am., Vol. 52, p. 1323, 1962.

> The output from one laser has been used to switch off laser oscillations in a second laser.

372. Mode Suppression and Single Frequency Operation in Gaseous Optical Masers. H. Kogelnik and C. K. Patel, Proc. IRE, Vol. 50, pp. 2365-2366, November 1962.

> The suppression of unwanted longitudinal modes in an inherently multimode He-Ne maser oscillation has been observed.

373. Visual Display of Isolated Optical Resonator Modes. H. Kogelnik and W. W. Rigrod, Proc. IRE, Vol. 50, p. 220, February 1962.

> A technique has been found for isolating a number of pure modes of a concave spherical cavity. The external cavity is adjusted to a nearly spherical condition, limiting the number of modes that can be excited. Then by adjusting maser gain and manipulating the mirrors and windows to utilize irregularities in their optical properties it is possible to raise the losses for all but one mode above the threshold for oscillation.

374. Measurements of the Laser Output. S. Koozekanani, Proc. IRE, Vol. 50, p. 207, 1962.

> Quantitative measurements of the output energy with a

simulated blackbody absorber have been made. In another experiment the blackbody absorber was used to determine the output of a 1-meter-long helium-neon gas laser. This measurement indicated an output of 3 mw.

375. Observations of Quasi-CW Operation of an Optical Ruby Maser. S. Koozekanani, M. Ciftan, and A. Krutchkoff, Appl. Optics, Vol. 1, pp. 372-373, May 1962.

 The results of experiments on ruby lasers is described. The experiments have led to the achievement of quasi-cw operation whereby all the spikes in the ruby laser output have been eliminated.

376. Measurements of the Laser Output. S. Koozekanani, P. P. Debye, A. Krutchkoff, and M. Ciftan, Proc. IRE, Vol. 50, p. 207, February 1962.

 Laser output energy is quantitatively measured with a simulated blackbody absorber.

377. Probabilities for the Neon Laser Transitions. G. F. Koster and H. Statz, J. Appl. Phys., Vol. 32, pp. 2054-2055, October 1961.

 Approximate calculations of the transition probabilities are reported.

378. Theory of Laser Oscillations in Fabry-Perot Resonators. M. J. Kotik and M. C. Newstein, J. Appl. Phys., Vol. 32, pp. 178-186, February 1961.

 The oscillation condition for a Fabry-Perot laser is derived from an integral equation for the angular spectrum of the field. The kernel of the integral equation involves the scattering matrices of the end mirrors. This integral equation leads to a stationary expression. The use of physically reasonable trial spectra allows one to estimate the effect of "walk-off" diffraction, reflector curvature, and reflector tilt in terms of an effective reflection coefficient for the infinite-aperture Fabry-Perot resonator. An approximate condition for the oscillation is derived.

379. Proposed Pumping Scheme for Continuous Laser. J. Kremen, Appl. Optics, Vol. 1, pp. 773-774, November 1962.

The use of ellipsoidal parabolic and spherical mirrors is investigated. Methods for obtaining a one-to-one image of the source at the laser crystal are presented.

380. Parametric Amplification in Spatially Extended Media and Application to the Design of Tunable Oscillators at Optical Frequencies. N. M. Kroll, Phys. Rev., Vol. 127, pp. 1207-1211, August 1962.

A theory of traveling-wave and backward-wave variable-parameter amplification appropriate to the amplification of a light beam is developed. It is an extension of the theory of Tien and Suhls for one-dimensional propagation to the case on which the pump wave, signal wave, and idler waves have different directions of propagation.

381. Parametric Amplification in Spatially Extended Media and Application to the Design of Tunable Oscillators at Optical Frequencies. N. M. Kroll, Proc. IEEE, Vol. 51, pp. 110-114, January 1963.

A theory of traveling-wave and backward-wave variable-parameter amplification appropriate to the amplification of a light beam is developed. The theory is then applied to the design of a tunable oscillator at optical wavelengths. Tuning is accomplished by changing the orientation of a parallel mirror system. A tuning range of three-to-one in frequency may be possible.

382. Directionality Effects of GaAs Light-Emitting Diodes. Part II. R. A. Laff, W. P. Dumke, F. H. Dill, Jr., and G. Burns, IBM J. Res. Dev., Vol. 7, pp. 63-65, January 1963.

The directional effects associated with a rectangular GaAs diode which has a length-to-width ratio of 10:1 in the plane of the junction are described.

383. Threshold Relations and Diffraction Loss for Injection Lasers. G. J. Lasher, IBM J. Res. Dev., Vol. 7, pp. 58-61, January 1963.

Mathematical expressions are derived for the minimum current density necessary to cause stimulated emission in injection lasers. A new type of diffraction loss for a thin light-emitting layer surrounded by light-absorbing material is calculated.

384. Spontaneous and Stimulated Line Shapes in Semiconductor Lasers.

G. J. Lasher and Frank Stern, Bull. Am. Phys. Soc. II, Vol. 8, p. 201, 1963.

Calculations of the spontaneous and stimulated-emission processes assuming an optical interband matrix element independent of the energy of initial and final states have been carried out for various temperatures and electron and hole quasi-Fermi levels.

385. Coherent Infrared Radiations in the Range 8-10 Microns by Gaseous Lasers. P. Laures, Compagnie Générale de Télégraphie Sans Fil, Paris, France, Third International Symposium on Quantum Electronics, Paris, France, February 1963.

Two gaseous lasers are examined which can lead to new sources in the range 8-10 microns corresponding to a propagation window in the atmosphere.

386. Solid-State Devices Other Than Semiconductors. B. Lax and J. G. Mavroides, Proc. IRE, Vol. 50, pp. 1011-1024, May 1962.

A review of laser development is included.

387. Semiconductor Masers. B. Lax, Polytechnic Institute of Brooklyn Symposia Series, XIII, Optical Masers, April 1963.

The various schemes proposed for achieving maser action in semiconductors include a variety of phenomena involving transitions between impurity levels, interband transitions, and transitions between magnetic levels. The operation and theory of the electrically pumped p-n junction diode of GaAs and related compounds are described in detail. The current status of the cyclotron resonance, magneto-optical, and indirect transition semiconductor masers is reviewed.

388. A Simple Method for Calibration of Ruby Laser Output. R. C. C. Leite and S. P. S. Porto, Proc. IEEE, Vol. 51, pp. 607-608, 1963.

A simple method of measurement of ruby laser outputs by reflecting the laser light in a $BaSO_4$ diffuse reflector and measuring the diffused light in a photocoil is presented.

389. Optical Maser Action in Eu^{3+} Benzoylacetonate. A. Lempicki and H. Samelson, Bull. Am. Phys. Soc., Vol. 8, p. 380, April 1963.

Optical maser action has been observed in an alcohol solution of Eu-benzoylacetonate at temperatures between
–130 and –170°C. Laser emission is observed at 6132 A.

390. Lasers. B. A. Lengyel, 136 pp., John Wiley and Sons, New York,
February 1963.

The topics covered include an introduction to the physical
problem of light generation; the development of the quantitative general relations; a presentation of the quantitative
general relations; a presentation of quantitative physical
and technical data and results, and a discussion of their relation to the analytical results.

391. Calculation of Giant Pulse Formation in Lasers. B. A. Lengyel
and W. G. Wagner, Hughes Research Laboratory, Malibu, California, Third International Symposium on Quantum Electronics,
Paris, France, February 1963.

A solution to the differential equations governing the evolu-
tion of the giant pulse is obtained. The solution permits
exact calculation of the energy, the peak power radiated, and
the inversion remaining at the end of the pulse. From the
formalism developed, conclusions are drawn concerning the
expected variation of the giant pulse parameters with temperature and the geometrical parameters of the laser.

392. Lasers. A. K. Levine, Am. Scientist, Vol. 51, pp. 14-31, March
1963.

An excellent review of the development of the laser is
presented.

393. Observations on the Pump-Light Intensity Distribution of a Ruby
Optical Maser with Different Pumping Schemes. T. Li and S. D.
Sims, Proc. IRE, Vol. 50, pp. 464-465, April 1962.

Several interesting observations on the performance of a
pulsed ruby optical maser pumped by a linear flash lamp in
an elliptical cavity and by a helical flash lamp in a cylindrical diffused reflector cavity are described.

394. A Calorimeter for Energy Measurements of Optical Masers. T.
Li and S. D. Sims, Appl. Optics, Vol. 1, pp. 325-328, May 1962.

The calorimeter consists essentially of several thermistor
beads connected in a bridge circuit and placed in intimate

contact with a carbon cone into which the maser beam is directed. The instrument is wide-band and easily calibrated.

395. Laser Beam Induced Electron Emission. D. Lichtman and J. F. Ready, Phys. Rev. Lett., Vol. 10, pp. 342-344, April 1963.

Sharp pulses of electron emission have been observed emanating from the target used in studying beam-surface interaction in vacuum.

396. Ultraviolet Absorption Spectra in Ruby. A. Linz, Jr., and R. E. Newnham, Phys. Rev., Vol. 123, pp. 500-501, July 1961.

The optical properties of highly doped rubies have been investigated. The optical absorption of a group of ultraviolet crystal-field bands near 3400 A was studied as a function of temperature, crystal orientation, and chemical composition. The intensity of the absorption bands varies with the square of the Cr concentration.

397. Quasi-Continuous Output from a Ruby Optical Maser. M. S. Lipsett and L. Mandel, Nature, Vol. 197, pp. 547-548, February 1963.

Quasi-continuous output is reported at room temperature from a ruby optical maser oscillating with a very nonuniform distribution of energy along the length of the maser rod.

398. Spectral Analysis of Fluctuations in Superposed Light from Two Ruby Optical Masers. M. S. Lipsett and L. Mandel, Dept. of Physics, Imperial College, London, U. K., Third International Symposium on Quantum Electronics, Paris, France, February 1963.

Fluctuations resulting from the superposition of light beams from two independent ruby optical masers are compared with fluctuations in the individual beams.

399. Mode Control in Ruby Optical Maser by Means of Elastic Deformation. M.S. Lipsett and M. W. Strandberg, Appl. Optics, Vol. 1 pp. 343, 357, May 1962.

A new technique has been applied in experimental work on ruby optical masers with dramatic results. It involves elastically deforming the medium, thereby involving the stress optic effect.

400. Coherence and Radiation Characteristics of Ruby Laser. I. D. Liu

and N. S. Kapany, J. Opt. Soc. Am., Vol. 52, p. 594, May 1962.

> The degree of coherence of laser oscillations can be corre-
> lated to the visibility of the interference fringes in the clas-
> sical Young's two-slit experiment. The intensity of the
> maxima and minima in the interference pattern is measured
> photoelectrically using fiber optics probes. It has been found
> that the degree of coherence varies from one individual pulse
> to the other and from one region to the other in a laser output.

401. Water Vapor Absorption Studies with a Helium-Neon Optical Maser.
Ronald K. Long and T. H. Lewis, Antenna Laboratory, Ohio State
University, Columbus, Ohio, Lasers and Applications Symposium,
Ohio State University, November 1962.

> Absorption of 1.153-micron radiation from a helium-neon
> cw optical maser by water vapor has been studied using a
> long-path-type absorption cell. Measurements are presented
> as a function of water vapor partial pressure, total pressure,
> using nitrogen as a broadening gas, and path length. Absorp-
> tion cell path lengths up to one kilometer were used. Partic-
> ular emphasis is placed on low absorber concentrations,
> corresponding to high-altitude atmospheric paths.

402. Atmospheric Attenuation of Ruby Lasers. Ronald K. Long, Proc.
IEEE, Vol. 51, pp. 859-860, 1963.

> In the case of a ruby laser as a source for either radar or
> communications experiments in the atmosphere a number
> of propagation problems are potentially troublesome. Here
> the molecular absorption problem as it relates to ruby lasers
> is discussed. The absence of water vapor absorption at
> 6943 A has been confirmed at the Antenna Laboratory by
> measurement with the laser and a 1-km laboratory absorp-
> tion cell.

403. Optical Properties of Paramagnetic Solids. W. Low, pp. 410-427
in Quantum Electronics, C. H. Townes, ed., Columbia University
Press, New York, 1960.

> The main method and results of energy level calculations
> for iron and rare earth ions are summarized. Intensity and
> line widths are discussed.

404. Some Factors Affecting Applicability of Optical Band Radio (Coherent Light) to Communications. D. Luck, RCA Review, Vol. 22, pp.

normal359-409, September 1961.

Technical factors determining the suitability of optical-band transmission are described, including propagation effects, concentration of signals in frequency and space, noise, power capability, and Doppler effects. Characteristics of equipment for working with radio signals in the optical band are described, with particular attention to coherent optical power generation, power supply conversion devices, modulators, and detectors. Some system performance comparisons are made with older bands.

405. Design of a Helium-Neon Gaseous Optical Maser. C. F. Luck, R. A. Paananen, and H. Statz, Proc. IRE (Correspondence), Vol. 49, pp. 1954-1955, December 1961.

Construction information is provided for a gaseous optical maser.

406. Detection of Coherent Light by Heterodyne Techniques with Solid-State Photodiodes. G. Lucovsky, R. B. Emmons, B. Harned, and J. K. Powers, Philco Scientific Laboratory, Blue Bell, Pennsylvania, Third International Symposium on Quantum Electronics, Paris, France, February 1963.

Quantum efficiency, current collection efficiency, and power conversion efficiency are developed as parameters to describe the operating characteristics of solid-state photomixers. Experimental data relating to the performance of InSb, Ge, Si, and GaAs photodiodes are presented. Efficiency parameters are computed from the experimental data and are found to compare well with predicted performance.

407. Masers, Iraser, and Laser. H. Lyons, Astronautics, Vol. 5, pp. 38-39, 100-104, May 1960.

Reference is made to potassium and cesium lasers under development at Columbia University.

408. Rare Earth Chelates and the Molecular Approach to Lasers. H. Lyons and M. L. Bhaumik, Electro-Optical Systems, Pasadena, Calif., Third International Symposium on Quantum Electronics, Paris, France, February 1963.

Lasers may be roughly divided into two classes, atomic and molecular. Molecules provide a wide and almost continuous

range of wavelengths and physical and optical properties.
Chemical tuning should be possible. Rare earth chelates are
of particular interest because of the ease of pumping, high
quantum efficiency, and narrow emission widths.

409. Fluorescence of Europium Tungstate. R. E. MacDonald, M. J. Vogel,
and J. W. Brookman, IBM J. Res. Dev., Vol. 6, pp. 363-364, July
1962.

The preparation and fluorescence properties of the material
are reported.

410. Rotation Rate Sensing with Traveling-Wave Ring Lasers. W. M.
Macek and D. T. Davis, Jr., Appl. Phys. Lett., Vol. 2, p. 678,
February 1963.

Sensing of rotation rate with respect to an inertial frame of
reference has been demonstrated using a cw He-Ne gas traveling-wave ring laser.

411. Ring Laser Rotation Rate Sensor. W. M. Macek, D. T. Davis, Jr.,
R. W. Olthuis, J. R. Schneider, and G. R. White, Polytechnic
Institute of Brooklyn Symposia Series, XIII, Optical Masers, April
1963.

The sensing of rotation rate with respect to an inertial frame
of reference has been demonstrated using a traveling-wave
gas laser in which the light propagates around in a closed
loop. Under stationary conditions, a degenerate pair of independent modes exists which consists of waves propagating
in opposite directions around identical paths. Rotation of the
laser about an axis with a component normal to the plane of
the laser path results in an effective lengthening of one path
and shortening of the other.

412. Microwave Modulation of Light. W. M. Macek, R. Kroeger, and
J. R. Schneider, IRE Int. Conv. Rec., Vol. 10, pp. 158-176, March
1962, Part 3.

This paper contains a discussion of the advantages inherent
in utilizing modulated light beams for space communications
and navigation. Light source requirements are reviewed.
The need to extend the modulating frequencies beyond the 100
Mc region is mentioned, including the specific applicability
of the Pockels effect. Chief emphasis is devoted to the application of the Pockels effect to an electro-optical microwave
light beam modulator.

413. An Interference Experiment with Two Independent Beams of Ruby Maser Light. G. Magyar and L. Mandel, Dept. of Physics, Imperial College, London, U. K., Third International Symposium on Quantum Electronics, Paris, France, February 1963.

> The transient interference fringes produced by two lasers are studied. The variation of the fringe visibility with exposure time gives information on the temporal coherence properties of the light.

414. Interference Fringes Produced by the Superposition of Two Independent Laser Beams. G. Magyar and L. Mandel, Nature, Vol. 198, pp. 255-256, April 1963.

> The observation of interference fringes produced by the superposition of two independent beams of ruby maser light is reported. The interference effects seem to be describable in completely classical terms.

415. Optical and Microwave-Optical Experiments in Ruby. T. H. Maiman, Phys. Rev. Lett., Vol. 4, pp. 564-566, June 1960.

> Experiments concerning the fluorescent relaxation process in ruby are reported. Also reported are the first observations of ground-state population changes in ruby due to optical excitation and the detection of optical absorption between two excited states in this crystal.

416. Optical Maser Action in Ruby. T. H. Maiman, Brit. Comm. and Electronics, Vol. 7, pp. 674-675, September 1960.

> The observation of optical oscillation in ruby (Cr_2O_3 in Al_2O_3) at a wavelength of 6943 A is reported.

417. Stimulated Optical Emission in Ruby. T. H. Maiman, J. Opt. Soc. Am., Vol. 50, p. 1134, November 1960.

> The fluorescence properties of ruby have been studied in detail. Negative temperatures and consequently amplification via stimulated emission is possible. An optical oscillator can be made using this principle which would serve as a very intense and coherent light source.

418. Stimulated Optical Radiation in Ruby. T. H. Maiman, Nature, Vol. 187, pp. 493-494, 1960.

An optical pumping technique has been successfully applied to a fluorescent solid resulting in the attainment of negative temperatures and stimulated optical emission at a wavelength of 6943 A. The active material used was ruby (chromium in corundum).

419. Stimulated Optical Emission in Fluorescent Solids, Part I; Theoretical Considerations. T. H. Maiman, Phys. Rev., Vol. 123, pp. 1145-1150, August 1961.

An analysis of stimulated emission processes in fluorescent solids is presented. The kinetic equations are discussed and expressions for pumping power and effective temperature of the exciting source are given in terms of the material parameters. A comparison of excitation intensity for three- and four-level systems is given. The spectral width of the stimulated radiation is discussed with particular attention to imperfect crystals.

420. Optical Maser Action in Ruby. T. H. Maiman, Advances in Quantum Electronics, J. R. Singer, ed., Columbia University Press, New York, 1961.

The results of an analysis of stimulated optical emission in fluorescent solids are presented. The results are applied to the case of ruby and compared with experimental data. Measurements were made on a 1.25 cm^3 sample under pulsed optical excitation. A peak power output of 5 KW, output energy near one joule, beam angle of 10^{-2} radian and spectral width of individual components in the output radiation of 0.6×10^{-3} A at 6943 A were measured.

421. State of the Art — Devices. T. H. Maiman, Polytechnic Institute of Brooklyn Symposia Series, XIII, Optical Masers, April 1963.

The state of the art is reviewed and recent advances are summarized.

422. Stimulated Optical Emission in Fluorescent Solids. T. H. Maiman, R. H. Hoskins, I. J. d'Haenens, C. K. Asawa, and V. Evtuhov, Phys. Rev., Vol. 123, pp. 1151-1157, August 1961.

Optical absorption cross sections and the fluorescent quantum efficiency in ruby have been determined. Stimulated emission from ruby under pulsed excitation has been observed to depend strongly on crystal perfection. It is suggested that mode instabilities due to temperature shifts and a time-varying magnetic field are contributing to an oscillatory behavior of the output pulse.

423. Effects of Dispersion and Focusing on the Production of Optical Harmonics. P. D. Maker, R. W. Terhune, M. Nisenoff, and C. M. Savage, Phys. Rev. Lett., Vol. 8, pp. 21-22, January 1962.

Dispersive effects which limit effective crystal thickness have been demonstrated in quartz. Effects of dispersion have been balanced out in potassium dihydrogen phosphate (KDP).

424. Optical Third Harmonic Generation in Various Solids and Liquids. P. D. Maker, R. W. Terhune, and C. M. Savage, Ford Motor Company, Dearborn, Michigan, Lasers and Applications Symposium, Ohio State University, November 1962.

The progress on studying optical third harmonic generation in a wide range of materials is described. In order to enhance the effect a giant pulse ruby laser with an output pulse of 0.2 joules in 30 nanoseconds was used. Using this laser a large increase in tripling under velocity-matched conditions in calcite (over previously reported results by the authors) was observed. As much as 10^{10} third harmonic photons per laser flash with the crystal in a focused beam was obtained.

425. On the Problem of Pulsed Oscillations in Ruby Maser. G. Makhov, J. Appl. Phys., Vol. 33, pp. 202-204, January 1962.

It is shown by means of a semiquantitative nonlinear analysis that elementary interaction between an inverted electron spin system and a resonant cavity does not give rise to the pulsed mode of operation of the ruby maser oscillator. It is suggested the additional nonlinearity necessary for the existence of such a mode resides in the "distant ENDOR" the interaction between the chromium electrons and the Al nuclei.

426. Dynamic Behavior of Quantum Mechanical Oscillators. G. Makhov and O. Risgin, The University of Michigan, Institute of Science and Technology, Ann Arbor, Mich., Third International Symposium on Quantum Electronics, Paris, France, February 1963.

Three distinct modes of operation occur in maser oscillators: overdamped, oscillatory transient, and recurrent sharp pulses of nearly equal amplitude. To establish conditions for the three types of behavior, rate equations of the three-level maser have been analyzed by semiquantitative methods of nonlinear mechanics. To account for the pulsed mode the population difference equation has been modified.

427. Observations on Pulse Structure in Ruby Laser Output. W. R. Mallory, Proc. IEEE, Vol. 51, p. 850, 1963.

    Observations of the output from a pulsed ruby laser are reported which indicate the presence of higher-frequency modulation of the output than the typical 1-microsecond spikes.

428. High Repetition Rate Pulsed Lasers. W. R. Mallory and K. F. Tittel, Polytechnic Institute of Brooklyn Symposia Series, XIII, Optical Masers, April 1963.

    The development of a low-threshold, high-repetition rate, 1-KW pulsed laser device using neodymium-doped materials is viewed. Investigations of efficiency, pump source, power supply, cooling, and component reliability are presented.

429. Some Coherence Properties of Non-Gaussian Light. L. Mandel, Department of Physics, Imperial College, London, U. K., Third International Symposium on Quantum Electronics, Paris, France, February 1963.

    The usual coherence properties of maser light have implications for the development of optical coherence theory. Maser light cannot be represented as a fluctuating wave with Gaussian amplitude distribution. Mutual coherence functions and spectral densities are second-order moments and not sufficient for a full description of the non-Gaussian light of an optical maser.

430. The Measures of Bandwidth and Coherence Time in Optics. L. Mandel and E. Wolf, Proc. Phys. Soc., Vol. 80, pp. 894-897, October 1962.

    Several different definitions of coherence time are investigated, and it is shown that the reciprocity relation $\Delta\tau \approx 1/\Delta\nu$ must be treated with caution, especially in the case of multiple-peaked distributions such as arise in optical masers.

431. Use of Electro-Optical Shutters to Stabilize Ruby Laser Operation. F. R. Marshall and D. L. Roberts, Proc. IRE, Vol. 50, p. 2108, October 1962.

> A Kerr electro-optical shutter has been used as the control element in a feedback loop to stabilize the operation of a ruby laser.

432. Energy Storage and Radiation Emission from Kerr-Cell-Controlled Lasers. F. R. Marshall, D. L. Roberts, and R. F. Wuerker, Bull. Am. Phys. Soc., II, Vol. 7, p. 445, August 1962.

> A description of the system used is given. Single pulses having a half-power width of 10 microseconds with peaks in excess of 50 mw have been generated.

433. Time Development of the Beam from the Ruby Laser. R. L. Martin, J. Opt. Soc. Am., Vol. 51, p. 477, April 1961.

> High-speed streak moving pictures of the beam from a ruby laser are discussed.

434. Coupling of Laser Rods. J. I. Masters, Proc. IRE, Vol. 50, pp. 220-221, February 1962.

> A geometry suitable for the coupling of laser rods is described. The coupled rods are used in the comparison of radiation in an interference experiment.

435. Propagation in Laser Crystals. J. I. Masters and G. B. Parrent, Jr., Proc. IRE, Vol. 50, pp. 230-231, February 1962.

> Experimental data on laser rods indicate an internal multiple reflection field characterized by a coherence function.

436. Laser Q Spoiling Effects Using a Remote Reflector. J. I. Masters and J. H. Ward, Proc. IEEE (Correspondence), Vol. 51, pp. 221-223, January 1963.

> This note describes the effect observed when an active rod that is totally reflecting at one end is provided with a carefully aligned remote but fixed mirror as the opposing reflector of the laser enclosure. The mechanism that appears to be effective is a simultaneous increase with mirror displacement of both the Q of the enclosure and its preoscillation losses.

437. Operation of a Nd^{3+} Glass Optical Maser at 9180 A. R. D. Maurer, Appl. Optics (Correspondence), Vol. 2, pp. 87-88, January 1963.

> The operation of a glass optical maser is described. A transition between crystal field components on the $4F_{3/2}$ state and the ground state is used.

438. Fluorescence and Stimulated Emission in Oxide Glass. R. D. Maurer, Polytechnic Institute of Brooklyn Symposia Series, XIII, Optical Masers, April 1963.

> Oxide glasses allow a variation of fluorescence properties by varying the base glass. A single soda-lime glass containing neodymium has been chosen for an intensive study. The fluorescence spectrum of this glass is described and contrasted with those of crystals. Quantitative calculations of pulse thresholds from measured fluorescence data show reasonable agreement. Spectral output characteristics will be given with those at threshold showing the multimode nature of single "spikes."

439. Effects of the Beam of a Ruby Laser on the Refraction Index of Some Substances. G. Mayer, and F. Gires, Compagnie Générale de Télégraphie Sans Fil, Corbeville S.-O., France, Third International Symposium on Quantum Electronics, Paris, France, February 1963.

> An instrumental device is described which was used to observe effects of intense beams of red light on the refractive indices n_i of quartz, PO_4H_2K, triglycine sulfate, and nitrobenzene.

440. Characteristics of Giant Optical Pulsations from Ruby. F. J. McClung and R. W. Hellwarth, Proc. IEEE, Vol. 51, pp. 46-53, January 1963.

> A method of laser modulation is described which produces fast, intense, and controllable giant laser pulses by Q modulation. A polarizer-Kerr cell combination placed inside the cavity causes the regeneration to switch between very high and very low losses.

441. Giant Optical Pulsations from Ruby. F. J. McClung and R. W. Hellwarth, J. Appl. Phys., Vol. 33, pp. 828-829, March 1962.

Giant pulses of optical maser radiation have been produced which are several orders of magnitude larger than the commonly observed spontaneous pulses. The pulses were produced by varying the effective reflectivity of the reflecting surfaces at the ends of the ruby rod through a Kerr-cell switching technique. The measured pulse characteristics are found to be in agreement with the theoretical predictions.

442. Optical Maser Action in the R_2 Line in Ruby. F. J. McClung, S. E. Schwarz, R. W. Hellwarth, I. J. d'Haenens, and F. J. Meyers, Bull. Am. Phys. Soc., II, Vol. 6, p. 511, December 1961.

By suppressing the R_1 line at high inversions maser action at the R_2 line in ruby is achieved, using endplates of multiple-layer dielectric interference filters.

443. R_2 Line Optical Maser Action in Ruby. F. J. McClung, S. E. Schwarz, and F. J. Meyers, J. Appl. Phys., Vol. 33, pp. 3139-3140, October 1962.

Because of a fast thermal relaxation it is not usually possible to observe stimulated emission at the R_2 line when stimulated emission is occurring at the R_1 line. R_1 emission is suppressed using multilayer dielectric interference mirrors for end reflectors.

444. Optical Spectra of Divalent Rate Earth Ions in Crystals. D. S. McClure and Z. Kiss, Polytechnic Institute of Brooklyn Symposia Series, XIII, Optical Masers, April 1963.

The rare earth ions may exist in the divalent state in suitable host crystals such as CaF_2. All of the trivalent ions from Ce to Yb are reduced in situ to the divalent state in CaF_2 by gamma irradiation. The spectra of these ions derive from f^n configurations, but the weak absorption due to these is masked at higher energies by strong broad bands of the parity-permitted $f^n: f^{n-1}d$ transitions.

445. Operation of a Gaseous Optical Maser at Wavelengths Out to 18.5 Microns and Oscillation on f-d Transitions in Neon. R. A. McFarlane, W. L. Faust, C. K. Patel, and C. G. Garrett, Bell Telephone Laboratories, Murray Hill, N. J., Third International Symposium on Quantum Electronics, Paris, France, February 1963.

Thirty-six optical maser oscillations at various wavelengths between 2.2 and 18.5 microns in radio-frequency excited discharges of neon, krypton, and xenon either alone or with the addition of helium are reported.

446. Oscillation on f-d Transitions in Neon in a Gas Optical Maser. R. A. McFarlane, W. L. Faust, and C. K. Patel, Proc. IEEE, Vol. 51, p. 468, March 1963.

The observation of oscillation at wavelengths in the 1.8-micron region on f-d transitions in neon not previously known to be inverted is reported.

447. New Helium-Neon Optical Maser Transitions. R. A. McFarlane, C. K. Patel, W. R. Bennett, Jr., and W. L. Faust, Proc. IRE, Vol. 50, pp. 2111-2112, October 1962.

The authors' optical maser investigations in rf discharges containing mixtures of helium and neon are summarized.

448. The Focusing of Light by a Dielectric Rod. J. McKenna, Appl. Optics,Vol. 2, pp. 303-310, March 1963.

The focusing effect of a dielectric circular cylinder in a uniform three-dimensional light field is studied. The cylinder consists of an absorbing core surrounded by a coaxial nonabsorbing sheath. The dielectric constants of the core and the sheath are the same. An expression for the energy density in the core is obtained.

449. Investigation of Ruby Optical Maser Characteristics Using Microwave Phototubes. B. J. McMurtry, Stanford Electronics Laboratories, Tech. Report 177-3, July 1962.

This report is concerned with the photoelectric mixing of optical signals separated by a microwave frequency. Successful operation of the microwave phototube is reported. The application of this method to the study of the ruby laser is discussed.

450. Investigation of Ruby Optical Maser Characteristics by Photoelectric Mixing Techniques. B. J. McMurtry, Sylvania Microwave Device Division, Mountain View, Calif., Third International Symposium on Quantum Electronics, Paris, France, February 1963.

The experiments permit conclusions concerning the behavior

of the laser's spectral output as a function of pump power and rod temperature, the validity of applying homogeneous interferometer mode analysis to the case of ruby, and the effects of fluorescence line shape and stimulated emission in determining oscillation frequencies.

451. Modulation and Direct Demodulation of Coherent and Incoherent Light at a Microwave Frequency. B. J. McMurtry and S. E. Harris, Stanford Electronics Laboratories, Tech. Report 176-3, September 1962.

The modulation and demodulation of both the incoherent light from a mercury-arc lamp and the coherent light from a ruby laser at a modulation frequency of 2700 Mc are described.

452. Direct Observation of Microwave-Frequency Beats Due to Photo-mixing of Ruby Optical Maser Modes. B. J. McMurtry and A. E. Siegman, Stanford Electronics Laboratories, Tech. Report 177-1, August 1961.

The observation of microwave signals produced by photo-mixing of near-neighbor axial-mode components in the output spectrum of a ruby optical maser is reported.

453. Photomixing Experiments with a Ruby Optical Maser and a Travel-ing-Wave Microwave Phototube. B. J. McMurtry and A. E. Siegman, Appl. Optics, Vol. 1, pp. 51-53, January 1962.

A standard oxide-cathode S-band traveling-wave tube has been used as an improvised microwave phototube to study the coherent light output from a ruby optical maser.

454. Multimoding and Frequency-Pulling Experiments on a Ruby Optical Maser. B. J. McMurtry and A. E. Siegman, J. Opt. Soc. Am., Vol. 52, p. 594, May 1962.

Experiments using a traveling-wave microwave phototube to detect the presence of several frequency components in the laser output are described.

455. Photomixing Experiments with a Ruby Optical Maser and a Travel-ing-Wave Microwave Phototube. B. J. McMurtry and A. E. Siegman, Appl. Optics,Supplement 1, pp. 133-135, 1962.

A standard oxide-cathode S-band traveling-wave tube is used

98

as a microwave phototube. Beat frequencies between the third and seventh nearest-neighbor axial modes produce microwave outputs. The mode interval is 600 mcs.

456. The Pockels Effect of Hexamethylenetetramine. R. Q. McQuaid, Appl. Optics, Vol. 2, pp. 320-321, March 1963.

Measurements indicate that HMTA has electro-optical properties approximating those of KDP and more favorable than those of ZnS or ADP. References on the Pockels effect are included.

457. Theory of Semiconductor Maser of GaAs. A. L. McWhorter, H. J. Zeiger, and Benjamin Lax, J. Appl. Phys., Vol. 34, pp. 235-236, January 1963.

Maser action appears to occur in even TE or TM modes guided along the plane of the junction like the surface modes on a dielectric slab.

458. Some New Aspects for Laser Communications. G. K. Megla, Appl. Optics, Vol. 2, pp. 311-315, March 1963.

The information rate which can be carried by an optical wave is critically examined in view of the dual nature of the electromagnetic waves.

459. Angular Distribution of Infrared Radiation from Lasing GaAs Diodes. A. E. Michel and E. J. Walker, Polytechnic Institute of Brooklyn Symposia Series, XIII, Optical Masers, April 1963.

The angular distribution of infrared radiation emitted by lasing GaAs diodes has been studied using a photographic technique. The devices were constructed from rectangular parallelepipeds cut from (100) GaAs wafers. The patterns observed differed markedly from device to device and depend strongly on the current level. Various features of the patterns will be discussed.

460. Determination of the Active Region in Light-Emitting GaAs Diodes. A. E. Michel, E. J. Walker, and M. I. Nathan, IBM J. Res. Dev., Vol. 7, pp. 70-71, January 1963.

Stimulated emission has been observed recently in GaAs diodes. A question of particular interest is the spatial origin of the light. In this letter it is shown that in high-efficiency units the light clearly comes from the p-region.

461. Optically Efficient Ruby Laser Pump. P. A. Miles and H. E. Edgerton, J. Appl. Phys., Vol. 32, pp. 740-741, 1961

Several laser systems were built in which the minimum energy to produce a threshold coherent pulse at 20°C varied between 200 and 320 joules. While the input pulse length has some effect, the main reason for the experimental energies having been so much greater than this minimal figure until now is inefficient optical design.

462. Second-Harmonic Generation of the $CaWO_3:Nd^{3+}$ Laser Lines and Mixing of Ruby and $CaWO_3:Nd^{3+}$ Pulsed Lasers in Piezoelectric Crystals. R. C. Miller and A. Savage, Bull. Am. Phys. Soc., II, Vol. 7, p. 397, June 1962

KDP is found to be most effective in harmonic generation and mixing.

463. Harmonic Generation and Mixing of $CaWO_4:Nd^{3+}$ and Ruby Pulsed Laser Beams in Piezoelectric Crystals. R. C. Miller and A. Savage, Phys. Rev., Vol. 128, pp. 2175-2179, December 1962.

Second harmonic generation has been observed in a variety of selected crystals. KDP and ADP were the most efficient nonlinear crystals investigated.

464. Harmonic Generation and Mixing of $CaWO_4:Nd^{3+}$ and Ruby Pulsed Laser Beams in Piezoelectric Crystals. R. C. Miller, Bell Telephone Laboratories, Murray Hill, N.J., Third International Symposium on Quantum Electronics, Paris, France, February 1963.

Potassium dihydrogen phosphate and ammonium dihydrogen phosphate were found to be the most efficient nonlinear crystals investigated. Among the materials which are opaque to the ruby second harmonic but which showed second harmonic generation with a focused Nd-laser beam are ZnO, $PbTiO_3$, CdS, and GaP.

465. The Luminescence from Ruby Excited by Fast Electrons. E. W. Mitchell and P. D. Townsend, Proc. Phys. Soc., Vol. 81, pp. 12-14, January 1963.

The spectrum of the luminescence from ruby excited by 2-

Mev electrons has been compared with that excited by ultraviolet light. Two effects have been noted: an additional pair
of lines at 7165 and 7218 A and an asymmetric broadening on
the long-wavelength side of the R doublet.

466. Scattering of Very Intense Light. M. Mizushima, Bull. Am. Phys.
Soc., II, Vol. 7, p. 444, August 1962.

The conventional dispersion formula for the scattering cross
section of light is extended to include the case of a very-
high-intensity light. The resulting formula has an extra factor which produces an appreciable decrease of the cross
section compared to the conventional one as the intensity of
the light beam increases.

467. Threshold Current for p-n Junction Lasers. J. L. Moll and J. F.
Gibbons, IBM J. Res. Dev., Vol. 7, pp. 157-159, April 1963.

A method of including the absorptive effects associated with
incomplete population inversion in the calculation of threshold current is described.

468. Monochromatic Ruby Optical Maser. L. F. Mollenauer, F. F.
Imbusch, H. W. Moos, and A. L. Schawlow, Bull. Am. Phys. Soc.,
II, Vol. 7, p. 445, August 1962.

By using a sapphire-clad rod immersed in liquid nitrogen a
single sharp line has been obtained. Its width was determined to be less than 0.0025 cm^{-1} near threshold.

469. Optical Maser Action of Organic Species in Amorphous Media.
D. J. Morantz, Polytechnic Institute of Brooklyn Symposia Series,
XIII, Optical Masers, April 1963.

Types of optical maser action possible are outlined for single organic species and for energy transfer systems. The
energy levels available and parameters relevant to successful action are discussed. The supporting material may be
crystalline, organic or inorganic glass, or liquid. The
choice of matrix involves the orientation of the active species and the relative populations of the various states. Self-
orientation in a laser bell is postulated and the possibilities
of the use of a coherent beam in micromanipulation in a
chemical system are considered.

470. Stimulates Light Emission by Optical Pumping and by Energy

Transfer in Organic Molecules. D. J. Morantz, B. G. White, and A. J. Wright, Phys. Rev. Lett., Vol. 8, pp. 23-24, January 1962.

A system consisting of a rigid glass containing benzophenone and naphthalene was allowed to absorb from white light. After the initial luminescence subsides, sufficient excited benzophenone exists to transfer energy to the naphthalene and cause repeated stimulated emission.

471. Interference Fringes with Long Path Difference Using He-Ne Laser. T. Morokuma, J. Opt. Soc. Am., Vol. 53, pp. 394-395, March 1963.

Interference fringes are observed at optical path differences up to 9 meters on a Michelson interferometer.

472. The Ultraviolet Absorption Spectra of Ruby. C. S. Naiman, Polytechnic Institute of Brooklyn Symposia Series, XIII, Optical Masers, April 1963.

The ultraviolet absorption spectra of ruby are discussed in light of three sets of data: the pair spectra occurring at about 28,000 to 30,000 cm^{-1}; the energy levels recently uncovered by excited-state absorption experiments performed on the long-lived E level; and the so-called "forbidden charge transfer bands" from 40,000 to 50,000 cm^{-1}.

473. Effect of Growth Parameters on the Threshold $CaWO_4$:Nd Crystals. K. Nassau, Polytechnic Institute of Brooklyn Symposia Series, XIII, Optical Masers, April 1963.

Of the various rare-earth-activated optical maser materials, trivalent neodymium in calcium tungstate represents a particularly desirable combination. Neodymium is used because the large ground-state splitting permits operation at room temperature, and calcium tungstate for low thresholds leading to continuous operation at room temperature. The outline of the Czochralski growth, substitution, and perfection of calcium tungstate single crystals has been recently presented. The continuation of this work has led to the elucidation of some of the factors of importance for the growth of better crystals.

474. GaAs Injection Laser. M. I. Nathan, IBM, T. J. Watson Research Center, Yorktown Heights, N. Y., Third International Symposium

on Quantum Electronics, Paris, France, February 1963.

A summary of the state of development of the injection laser
is discussed. Topics included are: directionality and mode
selection, spectral distribution of stimulated emission,
stimulated emission from 1.0°K to 300°K, observations at
cw operation, the nature of the transition which gives rise
to stimulated emission, the spatial distribution of radiation,
and calculation of threshold conditions.

475. Recombination Radiation in GaAs by Optical and Electrical Injection. M. I. Nathan and Gerald Burns, Appl. Phys. Lett., Vol. 1,
p. 89, December 1962.

Evidence is presented which strongly indicates that the
emission seen in the diodes is due to recombination involving an acceptor center.

476. Recombination Radiation in GaAs. M. I. Nathan and Gerald Burns,
Phys. Rev., Vol. 129, pp. 125-128, January 1963.

Sharp line emission near the absorption edge due to recombination of electrons and holes in recently available high-
purity GaAs has been observed. Exciton emission is observed and the temperature dependence of this line is discussed. Below the energy gap a set of three lines is observed, separated from each other by a longitudinal optical
mode phonon energy. In crystals grown in an O_2 atmosphere
an emission is observed apparently due to a recombination
of a bound exciton.

477. Stimulated Emission of Radiation from GaAs p-n Junctions.
M. I. Nathan, W. P. Dumke, G. Burns, F. H. Dill, Jr., and G.
Lasher, Appl. Phys. Lett., Vol. 1, pp. 62-64, November 1962.

The observation of the narrowing of an emission line from
a forward-biased GaAs p-n junction is reported. As injection current is increased the emission line at 77°K narrows
by a factor of more than 20 to a width of less than KT/5.

478. Solid State Research of the Applied Physics Department for the
Year 1961. Naval Ordnance Laboratory, AD290619, July 1962.

Included is a section on the study of laser mechanisms using
high-speed photography.

479. Control of Ruby Laser Oscillation by an Inhomogeneous Magnetic Field. H. C. Nedderman, Y. C. Kiang, and F. C. Unterleitner, Proc. IRE, Vol. 50, pp. 1687-1688, July 1962.

The control of the oscillations by the use of an inhomogeneous magnetic field is reported.

480. A Continuously Operating Ruby Optical Maser. D. F. Nelson and W. S. Boyle, Appl. Optics, Vol. 1, pp. 181-183, March 1962.

Cw operation of a ruby optical maser at 6934 A is reported. The pumping geometry consists of a short-arc high-pressure mercury lamp imaged upon the large end of a trumpet-shaped crystal immersed in liquid nitrogen.

481. The Polarization of the Output from a Ruby Optical Maser. D. F. Nelson and R. J. Collins, pp. 79-82 in Advances in Quantum Electronics, J. R. Singer, ed., Columbia University Press, New York, 1961.

Measurements of the polarization for ruby rods having several crystalline orientations are reported. Measurements are also made as a function of temperature from 100 to 300°K.

482. Coherence Experiments with a Pulsed Ruby Optical Laser. D. F. Nelson, R. J. Collins, and A. L. Schawlow, Bull. Am. Phys. Soc., II, Vol. 6, p. 68, February 1961.

The coherence of the emission from a pulsed ruby optical maser was measured using the two-slit interference patterns.

483. Measurement of Factors Affecting Threshold of a Continuously Operating Ruby Optical Maser. D. F. Nelson and D. E. McCumber, Bell Telephone Laboratories, Murray Hill, N.J., Third International Symposium on Quantum Electronics, Paris, France, February 1963.

The model of a continuous ruby laser proposed by Nelson and Boyle (Appl. Optics, Vol. 1, p. 181, 1962) has been generalized to include lifetime shortening due to stimulated emission, pump absorption of the Cr^{3+} ions in the metastable levels, pump depletion resulting from absorption and the consequent spatial variation along the crystal of the populations of the metastable levels.

484. Coherence Time of a Maser. H. E. Neugebauer, Appl. Optics, Supplement 1, pp. 90-91, 1962.

> The application of the accepted theory of coherence to optical masers is discussed.

485. Excitation of the Nd^{3+} Fluorescence in $CaWO_4$ by Recombination Radiation in GaAs. Roger Newman, J. Appl. Phys. (Communication), Vol. 34, p. 437, February 1963.

> This note reports the excitation of the 1.06-micron Nd ion fluorescence in $CaWO_4$ by recombination radiation from a GaAs p-n junction.

486. Generation of Laser Axial Mode Difference Frequencies in a Nonlinear Dielectric. K. E. Niebuhr, Appl. Phys. Lett., Vol. 2, pp. 136-137, April 1963.

> The optical generation of a microwave difference frequency in a nonlinear dielectric is reported.

487. Cathode Ray Laser Pump Device. J. W. Ogland and W. E. Horn, J. Opt. Soc. Am., Vol. 52, pp. 602-603, May 1962.

> The approach taken in the design of a cathode ray tube for continuous laser pumping is discussed, including the important features for phosphor utilization and optical efficiency.

488. Gain-Bandwidth in Optical Maser Amplifiers and Oscillators. R. C. Ohlmann and R. D. Haun, Jr., J. Opt. Soc. Am., Vol. 51, p. 473, April 1961.

> An analysis is presented which is intended as a heuristic guide for consideration of regenerative gain and of bandwidth narrowing in optical maser amplifiers and oscillators. By starting from the net rate of creation of photons in each mode of the maser enclosure, an expression has been derived for the steady-state gain of an optical maser amplifier.

489. The Evidence of Phase Uniformity at the Ruby Laser End Surface. Akira Okaya, IBM Communications Systems Department, Bethesda, Maryland, Lasers and Applications Symposium, Ohio State University, November 1962.

> The interference patterns due to slits or holes were often

used as the evidence of coherency of light. Ordinarily these interference patterns are not sufficient to prove extremely high coherency of laser beams. However, the patterns are sensitive to the initial relative phase between the light sources which are the laser beam radiated from various positions of end surface. From our experimental results we found that the interference patterns due to these holes were kept constant and had no smearing, no matter how the positions of the holes at the surface are changed for every flash at random. We could conclude that the phases of the laser radiation over the end surface are constant.

490. Some Potentialities of Optical Masers. B. M. Oliver, Proc. IRE, Vol. 50, pp. 134-135, February 1962.

An introduction to the principles and possible applications of the optical maser is presented. The methods of generating coherent radiation, of focusing it, and of collimating it into tight beams are described. The use of lasers for communications is explored, and certain medical and other applications are suggested.

491. Sparkling Spots and Random Diffraction. B. M. Oliver, Proc. IEEE, Vol. 51, pp. 220-221, January 1963.

It is proposed that coherent light reflected by a diffusing surface produces a complex random but stationary diffraction pattern.

492. Synthetic Maser Ruby. R. D. Olt, Appl. Optics, Vol. 1, pp. 25-32, January 1962.

Synthetic ruby, which is single crystal aluminum oxide doped with chromium oxide, is being extensively used in several solid state devices including microwave masers, coherent light oscillators (lasers) and a new phonon-type acoustical maser. This paper describes the physical properties of synthetic ruby which are pertinent to its use in solid-state applications.

493. Optical Maser Crystals. R. D. Olt, J. Opt. Soc. Am., Vol. 52, p. 601, May 1962.

Technical characteristics of crystals successfully used in optical maser operation to date are summarized, including output wavelength characteristics and physical properties.

494. Principles of Optical Communication. K. W. Otten, Electro-Technology, Vol. 70, pp. 111-130, September 1962.

> The intent of the article is to introduce and explain the advantages and disadvantages that light frequencies offer as compared to radio frequencies. In particular, it is intended to show the theoretical limitations of the information-carrying capacity of light due to the quantum nature of the energy transfer.

495. Direct Measurement of Optical Cavity q. R. A. Paananen, Proc. IRE, Vol. 50, p. 2115, October 1962.

> A technique for the direct measurement of optical cavity q by means of photomixing the outputs of two lasers is described.

496. Resonant Amplification in a Gas Maser. R. A. Paananen, Proc. IRE, Vol. 50, pp. 2115-2116, October 1962.

> Measurements of optical gain as a function of input signal level are reported.

497. Very-High-Gain Gaseous (Xe-He) Optical Maser at 3.5 Microns. R. A. Paananen and D. L. Boeroff, Appl. Phys. Lett., Vol. 2, pp. 99-101, March 1963.

> A gaseous optical maser with 0.015 torr of Xenon and 1.5 torr of helium gas has been found to have a gain of over 50 dbm at 3.508 microns.

498. Zeeman Effects in Gaseous He-Ne Optical Masers. R. A. Paananen, C. L. Tang, and H. Statz, Proc. IEEE, Vol. 51, pp. 63-69, January 1963.

> Experimental and theoretical results of a detailed study of the Zeeman effects in an He-Ne laser are given. Attention is confined to the strongest maser emission line ($2S_2:2P_4$). Under weak excitation conditions the maser emission is a doublet of right and left circularly polarized waves. Under normal excitation conditions the maser could oscillate in at least three modes.

499. A Light Source Modulated at Microwave Frequencies. J. I. Pankove and J. E. Berkeyheiser, Proc. IRE, Vol. 50, pp. 1976-1977, September 1962.

It is verified that efficient generation of light modulation at microwave frequencies is possible.

500. Theoretical Considerations on Millimeter Wave Generation by Optical Frequency Mixing. R. H. Pantell and J. R. Fontana, Proc. IRE, Vol. 50, pp. 1796-1800, 1962.

> The generation of radiation by mixing optical maser signals is one possible method for closing the gap between microwaves and infrared. The conversion efficiency attainable with different types of nonlinear media is considered.

501. Quantum-Mechanical Description of Maser Action at Optical Frequencies. Yoh-Han Pao, J. Opt. Soc. Am., Vol. 52, pp. 871-878, August 1962.

> This paper consists of a theoretical discussion of maser action at optical frequencies with emphasis on the differences between stimulated emission and absorption and incoherent fields. Use is made of a geometrical representation of the equations of motion of the density matrix to illustrate the conditions for continuous-wave operation, amplification, production of large pulses, and for the appearance of relaxation oscillations.

502. Quantum-Mechanical Description of Maser Action at Optical Frequencies. Yoh-Han Pao, J. Opt. Soc. Am., Vol. 52, pp. 871-878, August 1962.

> A geometrical representation of the equations of motion for the components of the density matrix has been developed for visualizing the internal processes of laser systems. Conditions necessary for the production of recurrent pulses, amplification, and for continuous wave operation are discussed in terms of this representation.

503. Partially Coherent Processes in Quantum Electronics. Yoh-Han Pao, Bell Telephone Laboratories, Murray Hill, N. J., Third International Symposium on Quantum Electronics, Paris, France, February 1963.

> This paper reports on the derivation of information for the study of the relation of the partially coherent properties of macroscopic radiation fields at optical frequencies to the details of the quantum-mechanical processes which give rise to them. The equation of motion for the density matrix rep-

108

resenting the quantized radiation field and the matter field is derived for an optical maser system of finite extent but having a very dense distribution of states in certain spectral regions of interest.

504. Frequency Behavior of an Optical Maser. J. H. Parks and A. Szoke, Bull. Am. Phys. Soc., II, Vol. 8, p. 379, April 1963.

Earlier calculations by W. E. Lamb on the power dependence of the oscillation frequency of a gaseous optical maser have been extended to include the effect of a small traveling wave that may be superimposed on the standing wave in a Fabry-Perot resonator operating in a single mode.

505. Coherence Theory with Application to Laser Light. G. B. Parrent, Jr., and Thomas J. Skinner, Technical Operations, Inc., Lasers and Applications Symposium, Ohio State University, November 1962.

This paper briefly reviews two formulations of coherent theory due to Wolf and the ensemble theory due to Beron. Emphasis is given to the physical interpretation of the correlation functions in terms of the general statistics of the sources and fields.

506. Anomalous Dispersion Optical Modulator and Demodulator. P. Parzen, RCA Laboratories, Princeton, N. J., Third International Symposium on Quantum Electronics, Paris, France, February 1963.

Two new solid-state devices, the reflection resonance modulator and the reflection resonance detector, are described. Both operate similarly and use a dielectric crystal in the anomalous dispersion region. Optical powers of 10^{-4} watt with modulation bandwidths of thousands of megacycles are detectable and a modulation index of thirty percent with similar bandwidths is attained.

507. Optical Power Output in He-Ne and Pure Ne Masers. C. K. N. Patel, J. Appl. Phys.,Vol. 33, pp. 3194-3195, November 1962.

An experimental setup is described for measuring the variation of optical power output with gas mixtures in a gas maser oscillator. Experimental results are given for the case of the He-Ne optical maser. A pure neon maser oscillator has been operated for the first time and the results are given.

508. Gaseous Optical Masers. C. K. N. Patel, Bell Telephone Laboratories, Murray Hill, New Jersey, Lasers and Applications Symposium, Ohio State University, November 1962.

> This article summarizes the present knowledge of gaseous discharge optical masers. After a short description of general conditions necessary for obtaining population inversion and maser oscillation in gaseous discharge media, the presently used optical cavities and the three sources of excitation in the gas discharge are discussed.

509. Infrared Spectroscopy Using Stimulated Emission Techniques. C. K. N. Patel, W. R. Bennett, Jr., W. L. Faust, and R. A. McFarlane, Phys. Rev. Lett., Vol. 9, pp. 102-104, August 1962.

> Optical maser oscillation has been obtained using electron excitation in each of the noble gases and on a total of 14 transitions falling in the wavelength range of 1.5 to 2.2 microns.

510. Optical Maser Oscillation in Pure He, Ne, Ar, Kr, and Xe. C. K. N. Patel, W. R. Bennett, Jr., W. L. Faust, and R. A. McFarlane, Bull. Am. Phys. Soc., II, Vol. 7, pp. 444-445, August 1962.

> Using electron impact, continuous optical maser oscillations have been obtained on 14 lines.

511. High-Gain Gaseous (Xe-He) Optical Masers. C. K. N. Patel, W.L. Faust, and R. A. McFarlane, Appl. Phys. Lett., Vol. 1, pp. 84-85, December 1962.

> A gaseous maser medium capable of large optical amplification at 2.026 microns is reported. It is found that the addition of relatively large amounts of helium to xenon resulted in an enormous increase in optical gain.

512. High-Gain Medium for Gaseous Optical Masers. C. K. N. Patel, W. L. Faust, and R. A. McFarlane, Bell Telephone Laboratories, Murray Hill, N. J., Third International Symposium on Quantum Electronics, Paris, France, February 1963.

> A mixture of helium and xenon has been used to obtain high gain. Gain varies inversely as the diameter of the discharge tube. The excitation processes leading to this high gain involve electron impact alone. The presence of high pressure

of helium increases the electron density as well as the average electron temperature.

513. Experimental Study of the Oscillation Modes of Optical Masers. M. Pauthier, Laboratoire Central des Télécommunications, Paris, Third International Symposium on Quantum Electronics, Paris, France, February 1963.

Oscillation modes are studied using helium-neon optical masers. For diverse configurations of cavities, lowest-order modes have been separated. A relative figure of merit for the configuration is based on measurements of coherence and power output.

514. Lasers. M. Pauthier, Elec. Commun., Vol. 37, pp. 377-386, 1962.

Laser action has been described concisely and clearly.

515. Optical Orientation of Molecules. A. Peikara and S. Kielch, Laboratoires d'Etudes des Diélectriques, Institut de Physique de l'Academie Polonaise des Sciences, Université A. Michieqicz, Posnan, Poland, Third International Symposium on Quantum Electronics, Paris, France, February 1963.

The high-amplitude electric field in the laser beam is sufficient to orient anisotropic molecules in some gases or liquids. This may involve observable changes in corresponding macroscopic properties such as the dielectric constant, refractive index, Rayleigh constant, degree of depolarization D, etc.

516. Nonlinear Optical Properties of Solids. P. S. Pershan, Division of Engineering and Applied Physics, Harvard University, Lasers and Applications Symposium, Ohio State University, November 1962.

The nonlinear optical properties of matter will be introduced into Maxwell's equations from a phenomenological point of view. For the simplest nonlinearity the existence of a time-averaged free energy will be shown to follow directly from the assumption of a nondissipative medium. The symmetry of the tension χ the describes the nonlinearity is derived from the free energy. Assuming a free energy, quadrupole and magnetic dipole nonlinearities will be discussed.

111

517. X-Band Microwave Phototube for Demodulation of Laser Beams. M. D. Petroff, H. A. Spetzler, and E. K. Bjornerud, Proc. IEEE, Vol. 51, pp. 614-615, April 1963.

> The authors describe their first results with an X-band microwave phototube utilizing an S-1 photocathode used together with a broad-band microwave interacting structure of simple but effective design.

518. Proposal for an Infrared Maser Dependent on Vibrational Excitation. J. C. Polanyi, J. Chm. Physics, Vol. 34, pp. 347-348, 1961.

> An infrared maser is proposed, involving inversion between vibrational states. Evidence is summarized that chemical reaction can lead directly to complete population inversions among freshly formed products of exothermic gaseous reactions.

519. Gaseous Optical Maser with External Mirrors. T. G. Polyanyi and W. R. Watson, J. Appl. Phys., Vol. 34, pp. 553-559, March 1963.

> Radiation patterns obtained with He-Ne masers are described. Various configurations of discharge tube ends are used. Mode selection by interposing objects in the cavity is described. Comparisons of lasers operating with external and internal mirrors show that the lack of azimuthal symmetry of the radiation patterns is intrinsic to the spherical mirror geometry and not to the Brewster angle windows.

520. Ruby Optical Maser as a Raman Source. S. P. Porto and D. L. Wood, J. SMPTE, Vol. 52, pp. 251-252, March 1962.

> The successful use of the ruby optical maser for excitation of Raman spectra is described. The conditions of the experiment are discussed and the utility of the ruby maser for this purpose is discussed.

521. Remarks on the Theory of Nonlinear Dielectrics. P. J. Price, IBM, Watson Research Laboratory, Lasers and Applications Symposium, Ohio State University, November 1962.

> A discussion of the fundamental physical principles (rather than detailed quantum theory) which apply to nonlinear-optical-frequency polarization. Linear polarization of a solid

is governed by some universal laws: Kramers-Kronig relations, sum rules, Onsager relations, and parity principles. The equivalents of these for the nonlinear polarization observed at optical frequencies are discussed.

522. Trigonal Sites and 2.24 Micron Coherent Emission of O^{3+} in CaF_2. S. P. Porto and A. Yariv, J. Appl. Phys. Lett., Vol. 33, pp. 1620-1621, April 1962.

> 2.24 micron coherent emission was found in crystals in which trigonal sites outnumbered tetragonal sites by 10:1.

523. Optical Maser Action in $BaF_2:U^{3+}$. S. P. Porto and A. Yariv, Proc. IRE, Vol. 50, pp. 1542-1543, June 1962.

> Measurements of output frequency, threshold, absorption, and the lifetime of the metastable level are reported.

524. Excitation, Relaxation and Optical Maser Action at 2.407 Microns in $SrF_2:U^{3+}$. S. P. Porto and A. Yariv, Proc. IRE, Vol. 50, pp. 1543-1544, June 1962.

> The determination of the pertinent energy levels, relaxation times, and excitation conditions for the stimulated maser emission are reported.

525. Low-Lying Energy Levels and Comparison of Laser Action of Trivalent Uranium in CaF_2, SrF_2, and BaF_2. S. P. Porto and A. Yariv, Bell Telephone Laboratories, Murray Hill, N. J., Third International Symposium on Quantum Electronics, Paris, France, February 1963.

> Studies of the absorption and fluorescence spectra of these materials have been used to determine the position of the $^4I_{11/2}$ and $^4I_{9/2}$ multiplets. Each of the systems was found to emit stimulated radiation in a number of wavelengths between 2.2 and 2.7 microns. The observed frequencies of the laser emission, some of which have not been reported previously are consistent with the energy level assignments.

526. Threshold and Stability of a Simplified Model Optical Maser. E. J. Post, Appl. Optics, Vol. 1, pp. 165-168, March 1962.

> The population behavior of three-level optical masers can be described by four nonlinear differential equations. The stability of such a set of equations is investigated. Instabil-

ity may occur if there is a discrepancy between the threshold power and the power needed to maintain the steady state.

527. Impurity Effects in a He-Ne Laser. J. K. Powers and B. W. Harned, Proc. IEEE, Vol. 51, pp. 605-606, April 1963.

   The influence of condensable impurities on the output of a He-Ne laser operating at 1.15 microns is reported.

528. Large Alkali Metal and Alkaline Earth Tungstate and Molybdate Crystals for Resonance and Emission Studies. S. Preziosi, R. R. Soden, and L. G. Van Uitert, J. Appl. Phys., Vol. 33, p. 1893, May 1962.

   Preparation of the crystals is described.

529. Theory of Mixing Laser Beams in Solids. P. J. Price and E. Adler, Bull. Am. Phys. Soc., II, Vol. 7, pp. 329-330, April 1962.

   Nonlinear generation of polarization fields at sum and difference frequencies causes mixing of two laser beams. These sum and difference fields may be calculated for an unpolarized insulating crystal by extending the standard treatment of optical polarizability to the next order of perturbation theory. Some of the terms of the complete result correspond to a heuristically calculated polarization due to the Lorentz field produced by the linear polarization current and the magnetic field.

530. Semiconductor Maser of GaAs. T. M. Quist, R. H. Rediker, R. J. Keyes, W. E. Krag, B. Lax, A. L. McWhorter, and H. J. Zeiger, Appl. Phys. Lett., Vol. 1, pp. 91-92, December 1962.

   Coherent radiation is reported from GaAs diode at 77°K. Performance is greatly improved at 4.2°K.

531. Some Recent Work with an Optically Pumped Cesium Laser. R. Rabinowitz and S. Jacobs, TRG, Syosset, New York, Third International Symposium on Quantum Electronics, Paris, France, February 1963.

   Properties of Cs lasers at 3.20 microns as well as 7.18 microns are described. The narrow Cs Doppler width makes possible the measurement of the hyperfine splitting of the Cs $8P_{1/2}$ state by tuning of the optical cavity.

114

532. Continuous Optically Pumped Cs Laser. R. Rabinowitz, S. Jacobs, and G. Gould, Appl. Optics, Vol. 1, pp. 513-516, July 1962.

A continuously operated 7.18-micron laser oscillator has been built using optically pumped cesium vapor as the amplifying medium. The power of 50 microwatts is coupled out of the confocal resonator by means of a 45° BaF_2 pick-off window. The measured intensity distribution is in good agreement with that derived from the Boyd-Gordon expression for the lowest-order mode.

533. Remarks on Negative Absorption. S. G. Rautian and I. I. Sobelman, Optics and Spectroscopy, Vol. 10, pp. 65-66, January 1961.

This article considers the possibility of obtaining negative absorption electronic transitions of organic molecules by means of optical methods of excitation.

534. Time Coherence in Ruby Lasers. J. F. Ready, Proc. IRE (Correspondence), Vol. 51, pp. 1695-1696, July 1962.

Evidence is presented that the coherence time for laser emission is greater than $1/\Delta\nu$.

535. Correlation of Output Spikes from Different Portions of a Ruby Laser. J. F. Ready, Appl. Optics, Vol. 2, pp. 151-153, February 1963.

This note describes an experiment in which the oscillatory outputs from different sections of a ruby laser rod are correlated.

536. Effects Due to Absorption of Laser Radiation. J. F. Ready, J. Opt. Soc. Am., Vol. 53, p. 514, April 1963.

A theoretical model for the effects of absorption of a high-power pulse of laser radiation at an opaque surface has been developed. The quantities calculated include the amount of material vaporized and the impulse delivered to the absorbing surface.

537. Effect of Mirror Alignment in Laser Operation. J. F. Ready and D. L. Hardwick, Proc. IRE, Vol. 50, pp. 2483-2484, 1962.

Uncoated ruby rods were mounted between the mirrors of a Hilger and Watts Fabry-Perot interferometer with their axes approximately normal to the mirrors and excited by two U-shaped flash tubes. In this way the parallelism of the end mirrors could be varied without changing any of the other parameters of the laser system.

538. Optical Pumping of Masers Using Laser Output. J. F. Ready and D. Chen, Proc. IRE, Vol. 50, pp. 329-330, March 1962.

Observation of stimulated emission at 6.4 kMc and higher using a ruby laser pump source is predicted.

539. Theoretical Consideration of Optical Maser Radiation. 1. The Resonant Mode Structure of a Ruby Fabry-Perot Cavity. M. Resnikoff and Yoh-Han Pao, Bull. Am. Phys. Soc., II, Vol. 6, p. 44, November 1961.

In connection with the formulation of a quantum theory for optical maser action the electromagnetic resonant mode structure of ruby rods has been examined theoretically.

540. Theoretical Consideration of Optical Maser Radiation. 2. A Quantum-Mechanical Description of Stimulated Emission within a Multimode Cavity. M. Resnikoff and Y.-H. Pao, Bull. Am. Phys. Soc., II, Vol. 6, p. 414, November 1961.

Quantum-mechanical equations describing the interaction between the atomic and radiation systems within a multimode cavity are formulated and numerical solutions obtained for cases of interest.

541. The Use of a Laser as a Light Source for Photographic Light Scattering from Polymer Films. M. B. Rhodes, D. A. Reddy, and S. S. Stein, AD293290, June 1962.

A V_V-polarized scattering pattern from a 1-mil-thick medium-density polyethylene film obtained with a single flash of the laser of about 500 microseconds duration is compared with a similar picture for the same sample requiring 4.5-hour exposure with conventional apparatus.

542. Time-Resolved Spectroscopy of Ruby Laser Emissions. S. L. Ridgway, G. L. Clark, and C. M. York, J. Opt. Soc. Am., Vol. 53, pp. 700-702, 1963.

The time variations of Fabry-Perot interference fringes produced by the coherent light from a ruby laser have been

recorded with an image converter camera. The recorded
variations indicate that: (a) the mode of oscillation of the
ruby can change from one relaxation-oscillation burst of
emission, or laser spike, to another; (b) a monotonic pro-
gression of wavelengths frequently occurs in a sequence of
spikes; (c) the oscillation mode can switch within a single
spike; and (d) several modes of oscillation can occur simultaneously in a single spike.

543.   The Granularity of Scattered Optical Maser Light. J. D. Rigden
and E. I. Gordon, Proc. IRE, Vol. 50, pp. 2362-2368, November
1962.

Diffraction limiting of the angular spread of the laser beam
is found to give rise to the granularity of scattered optical
maser light.

544.   Interaction of Visible and Infrared Maser Transitions in He-Ne.
J. D. Rigden and A. D. White, Bell Telephone Laboratories, Murray Hill, N. J., Third International Symposium on Quantum Electronics, Paris, France, February 1963.

A nearly confocal helium-neon maser with external mirrors
has been made to oscillate simultaneously at 6328 A and at
six wavelengths in the infrared. For the first time, one optical maser transition has been used to inhibit a strong oscillation at different wavelengths, thereby allowing other
weaker transitions to exhibit maser action.

545.   Isolation of Axisymmetrical Optical-Resonator Modes. W. W.
Rigrod, Appl. Phys. Lett., Vol. 2, pp. 51-54, February 1963.

A series of axisymmetric modes of the concave mirror interferometer, forming the external cavity of a gas-optical maser,
have been isolated. The isolation mechanism cannot be explained by linear mode theory alone, as in each case the diffraction losses of the dominant mode can exceed the losses of
the extinguished lower-order modes. It is suggested that the
highest-order mode prevails in competition for inverted
population with all lower-order modes because of its greater
beam area.

546.   Isolation of Axisymmetrical Optical-Resonator Modes. W. W.
Rigrod, Bell Telephone Laboratories, Murray Hill, N. J., Third
International Symposium on Quantum Electronics, Paris, France,

February 1963.

See No. 545.

547. Gaseous Optical Maser with External Concave Mirrors. W. W. Rigrod, H. Kogelnik, D. J. Brangaccio, and D. R. Herriott, J. Appl. Phys., Vol. 33, pp. 743-744, February 1962.

A helium-neon optical maser has been operated successfully with concave mirrors external to the gas-discharge tube. The latter is sealed at both ends with optically flat windows at the Brewster angle to the beam axis. Mirror alignment is less critical than for plane mirrors. Measured beat frequencies between several pairs of concave resonator modes agree with computed values.

548. Diffraction Studies with Plane-Parallel Maser Interferometer. W. W. Rigrod and A. J. Rustako, Jr., J. Appl. Phys., Vol. 34, pp. 967-968, April 1963.

A He-Ne laser tube supports oscillations at 6328 A, 1.15 and 3.39 microns. Calculations of Fox and Li are verified by means of apertures inside the resonators. Patterns made with various obstacles in the resonator bear a blurred resemblance to those made with the same obstacles in the 1-cm beam outside the resonator.

549. Cross Relaxation and Concentration Effects in Ruby. R. W. Roberts, J. H. Burgess, and H. D. Tenney, Phys. Rev., Vol. 121, pp. 997-1000, February 1961.

Cross relaxation effects in ruby maser crystals are treated by introduction of a cross relaxation probability in the rate equations. Detailed solutions have been obtained for several specific processes and compared to recent experiments. It is shown that cross relaxation can improve maser performance even in the absence of impurity doping.

550. Scattering of Light by Light in a Nonlinear Medium. H. R. Robl, U. S. Army Research Office, Durham, North Carolina, Third International Symposium on Quantum Electronics, Paris, France, February 1963.

This paper is concerned with photon-photon scattering due to the nonlinear polarizability of a continuous isotropic and nondispersive dielectric medium. The presentation is based on the quantization of the electromagnetic field in the dielectric medium. Results are presented in terms of the probability for the production of photon pairs at the intersection of two light beams, and the emission of single photons at the intersection of three beams.

551. Modulation of Coherent Light. J. E. Rosenthal, Bull. Am. Phys. Soc., II, Vol. 6, p. 68, February 1961.

A light valve is discussed which uses carrier density modulation of light passing through an epitaxial semiconductor sheet.

552. Optics of Radiation from Optical Masers. J. E. Rosenthal, Bull. Am. Phys. Soc., II, Vol. 6, p. 298, April 1961.

The effect of the high electric intensity of the energy band scheme of the medium used in the optical medium is considered.

553. Quantum Optics of Coherent Radiation from Masers. J. E. Rosenthal, Bull. Am. Phys. Soc., II, Vol. 6, p. 365, June 1961.

The probability of excitation to the upper of two states in an atomic system 2 hν apart by the absorption of two photons each of energy hν as compared with the probability of the same excitation by absorption of one photon h(2ν) is considered.

554. Physical Optics of Coherent Radiation Systems. J. E. Rosenthal, Appl. Optics, Vol. 1, pp. 169-172, March 1962.

Optical systems for long-range communications and short focus, high photon concentrations utilizing the ruby optical maser are discussed. Conditions are formulated that must be fulfilled by the coherent light generator to make it practically useful. These include the duration of time coherence and the geometry of the uniphasal wavefronts.

555. Toroidal Ruby Lasers. D. Ross, Proc. IEEE, Vol. 51, pp. 468-469, March 1963.

Stimulated emission travels around the axis of the toroid under total reflection. Resonator Q is determined by the reflectivity of the walls. Mode selection is controlled by the cross-section geometry.

556. Quantum Effects and Noise in Optical Communications. M. Ross, Proc. IEEE, Vol. 51, pp. 602-604, April 1963.

> Various aspects of quantum effects and noise in optical communication systems are discussed.

557. Measurement of the Properties of Laser Crystals at the Submillimeter Wavelengths. R. F. Rowntree and W. S. C. Chang, Antenna Laboratory, Ohio State University, Columbus, Ohio, Lasers and Applications Symposium, Ohio State University, November 1962.

> Since many crystals have strong lattice absorption bands in the far infrared, one of the most important steps in developing a submillimeter maser is to measure the dielectric properties of laser materials. This paper describes the measurement of these laser materials at the wavelength range from 100 microns to 1 mm by means of a special far-infrared spectrometer using an interferometric modulator as the "order-sorter." The index of refraction, the extinction coefficient, and the dispersion at 300°K and 100°K of single crystal materials such as $CaWO_4$ and MgO are presented.

558. Power Aperture and the Laser. M. D. Rubin, Proc. IRE, Vol. 50, pp. 471-472, April 1962.

> It is shown that the surveillance of a given number of square degrees of coverage per second out to some range with a given input noise requires at least the same average power aperture product at optical as at microwave frequencies.

559. A Comparison of the Energy Output of Various Solid State Laser Materials. Warren Ruderman, Isomet Corporation, Palisades Park, New Jersey, Lasers and Applications Symposium, Ohio State University, November 1962.

> Calorimetric studies were conducted of the output beams of a number of solid-state laser materials. Experiments were carried out under constant experimental conditions for geometry, pumping, crystal size and reflective coatings. Tests were carried out at room and liquid nitrogen temperatures. Materials evaluated included ruby, neodymium-doped glass,

neodymium-doped calcium tungstate, neodymium-doped
strontium molybdate, uranium-doped calcium fluoride and
dysprosium-doped calcium fluoride.

560. The Operation of Optical Masers in Uniform Magnetic Fields.
W. A. Runciman, National Magnet Laboratory, MIT, Cambridge,
Mass., Third International Symposium on Quantum Electronics,
Paris, France, February 1963.

The general features of the Zeeman splitting of the energy
levels of free atoms and ions in crystals is described. The
advantages of operating lasers in uniform magnetic fields
are outlined. These include tunability of the emission and
the possibility of simultaneous operation at two or more frequencies, which will give rise to sum and difference frequencies when incident on a nonlinear dielectric.

561. High-Efficiency Optical Maser Parameters. S. J. Sage, Appl.
Optics, Vol. 1, pp. 173-179, March 1962.

A study is made of the characteristics of an optical maser
system to optimize available parameters other than the
maser material itself. As each parameter is optimized a
major reduction in threshold level is realized. Finally, a
very lightweight optical maser is built and tested.

562. Detection and Amplification of the Microwave Signal in Laser Light
by a Parametric Diode. S. Saito, K. Kurokawa, Y. Fuji, T. Kimura,
and Y. Uno, Proc. IRE, Vol. 50, pp. 2369-2370, 1962.

The detection and amplification of microwave signals in
laser light by a conventional traveling-wave tube with an
oxide-coated cathode utilizes the square-law characteristic
of the "external photoelectric effect" of the photocathode.
Here, a similar effect is discussed from the "internal photoelectric effect" point of view.

563. Detection and Amplification of the Microwave Signal in Laser Light
by a Parametric Diode. S. Saito, Polytechnic Institute of Brooklyn
Symposia Series, XIII, Optical Masers, April 1963.

The microwave beat components of the adjacent longitudinal
modes of laser resonators made of ruby and other materials
were detected and amplified at microwave frequencies and
also mixed into intermediate frequencies by silver-bonded

121

parametric diodes of germanium and silicon. As one application, the author proposes a laser-radar system, in which the microwave-modulated radar light can be received by the one parametric amplifier for the microwave radar.

564. Stimulated Processes in Some Organic Compounds. N. Samelson and A. Lempicki, Bull. Am. Phys. Soc., II, Vol. 8, p. 380, April 1963.

Experiments were carried out on Eu-dibenzoylmethide, a metalloorganic complex showing a fluorescence characteristic of the rare-earth compounds, and benzophenone, a pure organic with a phosphorescence spectrum showing vibrational structure. When contained in a Fabry-Perot resonator EuD_3 shows evidence of stimulated emission under flash excitation. A high-loss mechanism prevents laser action. Benzophenone shows evidence of the latter effect only.

565. Optical Maser Design. J. H. Sanders, Phys. Rev. Lett., Vol. 3, pp. 86-87, July 1959.

Electron impact excitation is suggested as a pumping scheme for a gas medium. The Fabry-Perot etalon is suggested as the maser cavity.

566. Optical Masers. J. H. Sanders, J. Brit. IRE, Vol. 24, pp. 365-372, November 1962.

This review of the present situation in the optical maser field discusses the principle of the maser and its extension to the infrared and visible regions of the spectrum. Various types of optical maser which have been successfully operated are described. The unique features of high spectral purity and narrow beam width are pointed out, and some present and future applications are discussed.

567. Measurements of Second Harmonic Generation of the Ruby Laser Line in Piezoelectric Crystals. A. Savage and R. C. Miller, Appl. Optics, Vol. 1, pp. 661-664, September 1962.

The relative efficiency of a number of materials which generate the second harmonid of the ruby laser line has been measured. KDP was the most efficient. There seems to be no general correlation between second harmonic intensities and the physical properties of the crystals.

568. Infrared and Optical Masers. A. L. Schawlow, pp. 553-563 in Quantum Electronics, C. H. Townes, ed., Columbia University Press, New York, 1960.

> The design and construction of the proposed optical maser is discussed.

569. Infrared and Optical Masers. A. L. Schawlow, Bell Labs. Record, Vol. 38, pp. 402-407, November 1960.

> The requirements for a maser amplifier are stated in terms of the population of excitation levels, and are contrasted to the requirements for a maser oscillator where feedback is accomplished by terminating the amplifier with reflecting end walls. The active medium is required to have a natural resonance corresponding to a quantum transition between two energy levels at the desired operating frequency.

570. Solid State Optical Masers. A. L. Schawlow, J. Opt. Soc. Am., Vol. 51, p. 472, April 1961.

> Geometrical and quantum-mechanical requirements for a solid-state optical maser are investigated. Optical maser action at 7009 and 7041 A in concentrated ruby and at 6943 A in dilute ruby is reported.

571. Optical Masers. A. L. Schawlow, Sci. Amer., Vol. 204, pp. 52-61, June 1961.

> The history of the laser and various types of lasers are discussed.

572. Fine Structure and Properties of Chromium Fluorescence in Aluminum and Magnesium Oxide. A. L. Schawlow, pp. 50-64 in Advances in Quantum Electronics, J. R. Singer, ed., Columbia University Press, New York, 1961.

> Results of the investigation of the sharp-line spectra of chromium ions in crystals from a quantum-electronic point of view are reported.

573. Monochromatic Solid State Optical Masers. A. L. Schawlow, Polytechnic Institute of Brooklyn Symposia Series, XIII, Optical Masers, April 1963.

Emission lines from ions in crystals are sometimes nearly as sharp as those in gases. Because of the large number of ions per unit volume, the solids can give much greater amplification per unit length. It is thus possible to construct an optical maser without using a sharply resonant cavity. With some sacrifice in directional discrimination, it is possible to make a solid-state optical maser whose output wavelength is insensitive to the resonator and is determined almost entirely by the optical line. A monochromatic pulsed laser has been constructed. Stability is two parts in 10^7.

574. Simultaneous Optical Maser Action in Two Ruby Satellite Lines. A. L. Schawlow and G. E. Devlin, Phys. Rev. Lett., Vol. 6, pp. 95-98, February 1961.

The observation of stimulated emission is reported at 7010 and 7040 A from transitions in red ruby which arise from exchange coupling between neighboring chromium ions.

575. Infrared and Optical Masers. A. L. Schawlow and C. H. Townes, Phys. Rev., Vol. 112, pp. 1940-1949, December 1958.

The extension of maser techniques to the infrared and optical regions is considered. It is shown that by using a resonant cavity of centimeter dimensions having many resonant modes, maser oscillations at these wavelengths can be achieved by pumping with reasonable amounts of incoherent light.

576. Photoelectric Energy Meter for Measuring Laser Output. E. Schiel, Proc. IEEE, Vol. 51, pp. 365-366, 1963.

The most promising methods for measuring the total energy of a light pulse emitted by a laser involve the principles of absorption by a blackbody (calorimeters) and the integration of the photocurrent of a light-sensitive device. Recording the photocurrent on a scope will reveal the fine structure (spikes) of a laser pulse. Though theoretically it may be possible, experimentally it is difficult and time-consuming to integrate over many spikes of the laser pulse. A direct energy measurement utilizing a self-integrating circuit is described.

578. Investigation of the Growth of Optical Crystals. J. B. Schroeder,

AD285107, 5 pp., July 1962.

Techniques capable of growing optical-quality crystals for
solid-state laser applications are described. The doped crystals grown include CaF_2, BaF_2, $PbMoO_4$, and $CaWO_4$.

579. Induced and Spontaneous Emission in a Coherent Field. I. R.
Senitsky, Phys. Rev., Vol. 111, pp. 3-11, July 1958.

The interaction between a coherently oscillating radiation
field and a number of similar atomic systems coupled to the
field through their electric dipole moments is analyzed for
the case of resonance between atomic system and field with
both the field and the molecules treated quantum-mechani-
cally.

580. Induced and Spontaneous Emission in a Coherent Field. II. I. R.
Senitsky, Phys. Rev., Vol. 115, pp. 227-237, July 1959.

The interaction between the electromagnetic field and a number of identical atomic systems, individually characterized
by an electric dipole moment and two energy levels, is analyzed for the case where the atomic systems are inside a
lossy cavity and exposed to a coherent driving field resonance being assumed between atomic system, cavity, and
driving field.

581. Induced and Spontaneous Emission in a Coherent Field. III. I. R.
Senitsky, Phys. Rev., Vol. 119, pp. 1807-1815, September 1960.

The theory developed in the first two articles of this series
dealing with the interaction between the electromagnetic field
in a cavity resonator and a number of two-level molecules is
generalized to include a Gaussian spread in the molecular
frequency. The center of the molecular frequency distribution coincides with the cavity resonant frequency. There is a
coherent driving field in the cavity at the same frequency and
the cavity loss is taken into account.

582. Infrared and Spontaneous Emission in a Coherent Field. IV. I. R.
Senitsky, Phys. Rev., Vol. 123, pp. 1525-1537, September 1961.

The perturbation restriction of previous articles is removed,
allowing large changes in the field, but the molecules are

still assumed to undergo a small change during the time under consideration. The justification for this type of analysis, involving the generalization of the conventional concepts of induced and spontaneous emission, the applicability to a molecular amplifier during the buildup period, and the reexamination of a calculation by Serber and Townes concerning the fundamental limits of molecular amplification, is discussed.

583. Induced and Spontaneous Emission in a Coherent Field. V. Theory of Molecular Beam Amplification. I. R. Senitsky, Phys. Rev., Vol. 127, pp. 1638-1647, September 1963.

The emission in a cavity and the steady-state conditions when a molecular beam traverses the cavity are analyzed. The induced emission energy when averaged depends only on the energy spectrum of the driving field. Results are then applied to a molecular beam analysis, and conditions for a steady state in the cavity are obtained.

584. Amplification by Resonance Saturation in Millimeter Wave Cavities. B. Senitsky and S. Cutler, Polytechnic Institute of Brooklyn Symposia Series, XIII, Optical Masers, April 1963.

The nonlinear properties of media which exhibit power absorption saturation have been used to obtain amplification of millimeter-wave radiation. Amplification has been obtained with traveling-wave structures at 3.7 mm in a crystalline solid (Fe-doped TiO_2) and 3.5 mm in a molecular gas ($HCl^{12}N^{15}$). An amplifier based on this principle is described which consists of a simple resonant cavity filled with $HCl^{12}N^{14}$ gas and yielding a signal gain of 8.5 db at 3.4 mm wavelength.

585. Piezoelectric Optical Maser Modulator. B. O. Seraphin, D. G. McCauley, and L. G. LaMarca, Polytechnic Institute of Brooklyn Symposia Series, XIII, Optical Masers, April 1963.

The performance of a Fabry-Perot-type light modulator is described in which the intensity of the interference pattern in reflection is modulated by a piezoelectric change in thickness of the quartz spacer.

586. On the Time-Resolved High-Resolution Spectroscopic Study of

the Emission from the Ruby Laser. M. Shimazu, I. Ogura, A. Hashimoto, and H. Sasaki, Polytechnic Institute of Brooklyn Symposia Series, XIII, Optical Masers, April 1963.

>The fine structure of the emission from the ruby laser was observed by means of a Fabry-Perot interferometer. In order to study the time variation of the wavelength of these component lines, the time-resolved high-resolution spectrum was observed by using the ultra-high-speed streak camera combined with a Fabry-Perot interferometer. Scanning speed of the streak camera is 0.5 to 4 mm/microsecond.

587. Theory of Masers for Higher Frequencies. K. Shimoda, Sci. Papers Inst. Phys. Chem. Research (Tokyo), Vol. 55, pp. 1-6, March 1961.

>Some of the general problems encountered in the maser-type generation and amplification of infrared or optical radiation are discussed. Threshold conditions of oscillation and amplification for a beam maser, a solid-state maser, and a gas maser are obtained from a unified theory. Coherence of the high-frequency maser may be expressed in terms of the spectral width of the maser radiation.

588. Theory of Masers for Higher Frequencies and Its Application to a Microbeam Electronic Accelerator. K. Shimoda, J. Opt. Soc. Am., Vol. 51, pp. 472-473, April 1961.

>Some of the general problems encountered in the maser-type generation and amplification of infrared or optical radiation are discussed. A high-energy electron accelerator is proposed.

589. Threshold Condition of Masers for Higher Frequencies. Koichi Shimoda, Appl. Optics, Vol. 1, pp. 303-307, May 1962.

>Threshold conditions for beam, solid-state, and gas masers are calculated from a unified theory. Operation of a gas maser on the line broadened by both collision and the Doppler effect is discussed. The rate of producing population inversion is calculated in terms of gas pressure, quantum efficiency, saturation parameter and the intensity of excitation. Finally the optimum condition is discussed.

127

590. Proposal for an Electron Accelerator Using an Optical Maser. Koichi Shimoda, Appl. Optics, Vol. 1, pp. 33-35, January 1962.

A high-energy electron accelerator is proposed and discussed. This is one of many possible applications of extremely high brightness, temperature, and radiation density obtainable with the optical maser.

591. Amplitude and Frequency Variations in Optical Masers. Koichi Shimoda, Polytechnic Institute of Brooklyn Symposia Series, XIII, Optical Masers, April 1963.

The laser action of six ruby crystals is investigated in a variety of configurations. Fabry-Perot interferograms are investigated with a streak camera. A theory is given which explains the existence of continuous spiking. The condition of continuous spiking and spikeless oscillation is discussed.

592. Light Source System for Ruby Laser. G. Shinoda, T. Susuki, and M. Umeno, Japan, J. Appl. Phys., Vol. 1, pp. 364-365, December 1962.

A Xe flash lamp composed of two coaxial tubes is described. Its advantages include low discharge voltage (300 volts) and higher efficiency by a factor of four.

593. Nonlinear Optical Effects: An Optical Power Limiter. A. E. Siegman, Appl. Optics, Supplement 1, pp. 127-132, 1962.

A parametric subharmonic oscillator comprised of a Fabry-Perot resonator filled with nonlinear crystal and ends transparent at the pump frequency and reflecting at the subharmonic frequency will function as an ideal power limiter at the fundamental. Power transmission at the pump frequency will limit sharply at the subharmonic oscillation threshold. A large power-dependent reflection will also occur on the in-end above threshold.

594. Microwave Modulation and Demodulation of Light. A. E. Siegman, B. J. McMurtry, and S. E. Harris, Stanford Electronics Laboratories Tech. Report, 176-2, July 1962.

Methods for modulating and demodulating coherent light signals with modulation bandwidths in excess of 1000 Mc are considered. Emphasis is given to the potentialities of the microwave phototube.

595. Microwave Demodulation of Light. A. E. Siegman, B. J. McMurtry, and S. E. Harris, Stanford Electronics Laboratories, Stanford University, Calif., Third International Symposium on Quantum Electronics, Paris, France, February 1963.

Traveling-wave microwave phototubes, microwave semi-conductor diodes, FM-AM optical converters, FM discriminator microwave phototubes, and distributed-emission microwave phototubes are discussed.

596. Phonon Broadening of Narrow Line Spectra in Solids. G. H. Silsbee, Rockefeller Hall, Cornell University, Ithaca, New York.

In the case of narrow lines and for certain broadening mechanisms it is important to consider the relative magnitudes of the observed line breadth and the spread of frequencies associated with the lattice vibrations. If the line is narrow compared with this spread of frequencies a phenomenon similar to motional narrowing in magnetic resonance must be considered when calculating the line width.

597. Investigation of Photo Beats from Ruby Lasers. M. Silver, R. S. Witte, and C. M. York, Bull. Am. Phys. Soc., II, Vol. 8, p. 380, April 1963.

High-speed streak photography utilizing an STL image converter camera has been used to observe photo beats in the outputs of several ruby lasers.

598. Optical Design for Sun-Pumping a CW Optical Maser. G. R. Simpson, J. Opt. Soc. Am., Vol. 52, p. 595, May 1962.

An immersion optics system has been designed to enable end-pumping of an optical maser using the sun as a source.

599. Advances in Quantum Electronics. J. R. Singer, ed., Columbia University Press, New York, 1961.

This volume is a collection of the papers and discussions presented at the Second International Conference on Quantum Electronics, held at Berkeley, Calif., March 1961.

600. Optical Maser Utilizing Molecular Beams. J. R. Singer and

I. Gorog, Bull. Am. Phys. Soc., II, Vol. 7, p. 14, January 1962.

Molecular-beam optical masers provide continuous optical emission from molecules which can only be excited to an inverted Boltzmann distribution in an irreversible manner. A number of molecules including some alkali halides are examined as possibilities for obtaining optical maser action.

601. General Analysis of Optical, Infrared, and Microwave Maser Oscillator Emission. J. R. Singer and S. Wang, Phys. Rev. Lett., Vol. 6, pp. 351-354, April 1961.

The equations governing coherent emission from quantum mechanical amplifiers using either electric or magnetic dipole transitions are generalized. It is found that amplitude modulation of the output is to be expected from all maser amplifiers excepting those in which excited atoms are supplied at a notably higher rate than the depopulation rate due to coherent induced emission.

602. The Emission, Pulse Level Inversion, and Modulation of Optical Masers. J. R. Singer and S. Wang, pp. 299-307 in Advances in Quantum Electronics, J. R. Singer, ed., Columbia University Press, New York, 1961.

It is shown that for any two-level optical or microwave maser system, the interaction between the radiation field and the radiating molecules results in an amplitude modulation of the radiation output.

603. Blue Fluorescence in Crystals Excited by the Ruby Maser. S. Singh and B. P. Stoicheff, Polytechnic Institute of Brooklyn Symposia Series, XIII, Optical Masers, April 1963.

Intense blue fluorescence has been observed in several crystals: $Eu^{2+}:LaCl_3$, $Nd^{3+}:LaCl_3$, and $Nd^{3+}:LaBr_3$. The intensity of the blue fluorescence increases as the square of the maser beam intensity. The fluorescence decreases in intensity as the temperature is lowered, and is absent at $77°$ K.

604. An Analysis of the Maser Oscillator Equations. D. M. Sinnett, J. Appl. Phys., Vol. 33, pp. 1571-1578, April 1962.

This paper analyzes the maser oscillator equations which de-

scribe the interaction between the cavity and the inverted
population of the electron spin system of the paramagnetic
substance. It is shown that the equations will not allow
periodic solutions, thus refuting the theory based on computer solutions that this interaction is responsible for the
pulsed mode of operations of the oscillator. Characteristic
solutions of these equations are determined analytically.
Numerical solutions show that the periodic solutions may be
induced by supplementing the spin system equation with an
additional term.

605. Lasers for Aerospace Weaponry. Janis Sirons, Technical Docu-
mentary Report No. ASD-TDR-62-440, 42 pp., U. S. Department
of Commerce, OTS, May 1962.

This report presents a summary of the state of the art in
laser research. Characteristics of the present ruby laser
are described and compared with idealized characteristics.
Mathematical calculations of performance using existing
lasers and ways to improve existing lasers are described,
including new crystal materials, new sources of optical energy, and new lasing techniques.

606. An Active Interference Filter as an Optical Maser Amplifier.
V. N. Smiley, Proc. IEEE, Vol. 51, pp. 120-124, January 1963.

Approximate theoretical expressions for gain, bandwidth,
and root gain-bandwidth are derived by introducing negative
absorption into equations for a Fabry-Perot interference
filter. Root gain-bandwidth is shown to be a constant for
a given cavity as long as the cavity bandwidth is much small-
er than the Doppler or fluorescent linewidth of the maser
transition.

607. Experiments with a Long Gas-Phase Optical Maser. V. N. Smiley,
U. S. Navy Electronics Laboratory, San Diego, Calif., Third In-
ternational Symposium on Quantum Electronics, Paris, France,
February 1963.

Descriptions of an 8-meter tube, optical bench, type of cavity, and method of obtaining the discharge are given. Losses
in long Fabry-Perot and confocal cavities are discussed.
Methods of discharge excitation include with external electrodes and 60 cps ac and dc with internal electrodes. Values
of optical gain are higher than previously obtained. Beats

between axial modes separated by about 20 Mc are observed.

608. Optical Mixing of Coherent and Noncoherent Light. A. W. Smith and N. Braslau, IBM J. Res. Dev., Vol. 6, pp. 361-362, July 1962.

   The mixing of a ruby maser signal with spectral lines of a mercury lamp in a piezoelectric crystal is reported.

609. Optical Mixing of Coherent and Incoherent Light. A. W. Smith and N. Braslau, Polytechnic Institute of Brooklyn Symposia Series, XIII, Optical Masers, April 1963.

   The mixing of the ruby optical maser with spectral lines from a high-pressure mercury lamp will be discussed. Both sum and difference frequencies have been observed. Mixing takes place in crystals of KDP.

610. On the Detection of Maser Signals by Photoelectric Mixing. A. W. Smith and G. W. Williams, J. Opt. Soc. Am., Vol. 52, pp. 337-338, March 1962.

   A correction to the signal-to-noise ratios calculated in a previous article by A. T. Forrester (J. Opt. Soc. Am., Vol. 51, p. 253, 1961) is presented.

611. Project Lunar See. L. D. Smullin and G. Fiocco, Proc. IRE, Vol. 50, pp. 1703-1704, July 1962.

   Results of an optical maser radar experiment with the moon as a target are presented.

612. Optical Maser Action of Nd^{3+} in a Barium Crown Glass. E. Snitzer, Phys. Rev. Lett., Vol. 7, pp. 444-446, December 1961.

   A glass optical maser which operates at room temperature and contains trivalent neodymium as the active ion is described. The output consists of a number of sharp lines in a wavelength interval of about 30 A centered at approximately 1.06 microns.

613. Optical Dielectric Waveguides. E. Snitzer, pp. 348-369 in Advances in Quantum Electronics, J. R. Singer, ed., Columbia University Press, New York, 1961.

   By cladding a core of high index of refraction and of sufficiently small cross section with another material of lower

refractive index, dielectric waveguides can be made with
one or a few modes of propagation in the visible region of
the spectrum. The properties of glass fibers acting as waveguides are presented.

614. Neodymium Glass Optical Masers. E. Snitzer, J. Opt. Soc. Am.,
Vol. 52, p. 594, May 1962.

> Stimulated emission of trivalent neodymium has been obtained
> for concentration from 0.1 to 13 wt. % Nd_2O_3 in a barium
> crown glass. The emission consists of a series of sharp
> lines in a wavelength interval of from 5 to 70 A wide, cen-
> tered at 1.06 microns.

615. Neodymium-Glass-Fiber Laser. E. Snitzer, R. Crevier, and F.
Hoffman, J. Opt. Soc. Am., Vol. 53, p. 515, April 1963.

> The light transmission in fibers with diameters comparable
> to the wavelength of light is treated as a dielectric wave-
> guide.

616. Optical Waveguide Modes in Small Glass Fibers. E. Snitzer and
J. W. Hicks, J. Opt. Soc. Am., Vol. 49, p.1128, November, 1959.

> The light transmission in fibers with diameters comparable
> to the wavelength of light is treated as a dielectric waveguide.

617. Cylindrical Dielectric Waveguide Modes. E. Snitzer and H. Oster-
berg, J. Opt. Soc. Am., Vol. 51, pp. 499-505, May 1961.

> The propagation of cylindrical dielectric waveguide modes
> near cutoff and far from cutoff are considered. The relative
> amounts of E_Z and H_Z and the transverse components of the
> field are determined for both sets of hybrid modes.

618. High-Power Pulsed Neodymium Glass Laser. E. Snitzer, J. Opt.
Soc. Am., Vol. 52, p. 1323, 1962.

> An energy of 113 joules has been obtained from a 1/4-in.-
> diameter clad rod, 18 in. long. A straight flash tube was
> used with an electrical energy input of 9000 joules. The out-
> put pulse lasted for 1.6 msec, with a beam spread of some-
> what less than 10°.

619. Rate Equation Approach to the Performance of Q-Switched Lasers.
N. Solimene, TRG, Syosset, New York, Third International Symposium on Quantum Electronics, Paris, France, February 1963.

> The behavior of the laser is analyzed. Using a model based

on the phenomenological rate equations, population density of energy levels and the optical energy density in the resonator are assumed to be uniform. It is further assumed that during the time required for the development of the output pulse or pulses the population difference is affected only by the process of induced emission, while the sum of the populations of the upper and lower levels remains constant. Output energies and pulse shapes are computed.

620. Dynamics Limitations on the Attainable Inversion in Ruby Lasers. W. R. Sooy, R. S. Congleton, and W. K. Ng, Hughes Aircraft Co., Culver City, Calif., Third International Symposium on Quantum Electronics, Paris, France, February 1963.

Five different mechanisms which act to limit the inversion in the ruby are considered. These are: spontaneous decay, prelasering, lateral depumping, internal modes in the ruby, and superradiance. An analysis of these effects and the results of measurements of the inversion as a function of radius and the effects on inversion of surface roughening, immersion, and pumping geometry are presented.

621. Spectral Characteristics of CW GaAs Lasers. P. P. Sorokin, J. D. Axe, and J. R. Lankard, Polytechnic Institute of Brooklyn Symposia Series, XIII, Optical Masers, April 1963.

High-resolution measurements were made of the spectral distribution of the output beam of a cw GaAs laser. Narrowing of both primary and secondary components was observed. Oscillations in modes other than the lowest-loss ones were observed. From the spectra one can deduce the number of photons in each component, as well as the ratio of decay times for lowest-loss and next-to-lowest-loss modes.

622. Stimulated Infrared Emission from Trivalent Uranium. P. P. Sorokin and M. J. Stevenson, Phys. Rev. Lett., Vol. 5, pp. 557-559, December 1960.

The characteristics of stimulated emission from trivalent uranium ions substituted for divalent calcium ions in calcium fluoride are described.

623. Solid State Optical Masers Using Trivalent Uranium and Divalent Samarium. P. P. Sorokin and M. J. Stevenson, J. Opt. Soc. Am., Vol. 51, p. 477, April 1961.

Optical maser operation in calcium fluoride is reported. Pumping power required to reach threshold for oscillation is less by a factor of more than 500 than in the case of the ruby optical maser. In the uranium system emission occurs at 2.5 microns and in the samarium system at 7082 A.

624. Solid State Optical Maser Using Bivalent Samarium in Calcium Fluoride. P. P. Sorokin and M. J. Stevenson, IBM J. Res. Dev., Vol. 5, pp. 56-58, 1961.

Laser action is reported. The system may potentially be operated cw.

625. Stimulated Emission from $CaF_2:U^{3+}$ and $CaF_2:Sm^{2+}$. P. P. Sorokin and M. J. Stevenson, pp. 65-77 in Advances in Quantum Electronics, J. R. Singer, ed., Columbia University Press, New York, 1961.

The four-level maser system is analyzed. The essential characteristic of a four-level maser is that oscillation occurs in a transition between a metastable state and a terminal state located sufficiently far above the ground state to be virtually unoccupied, resulting in a much lower power requirement.

626. Solid State Optical Maser Using Divalent Samarium in Calcium Fluoride. P. P. Sorokin and M. J. Stevenson, IBM J. Res. Dev., Vol. 7, pp. 56-58, January 1963.

A new solid-state optical maser with an output in the red portion of the visible spectrum is reported. One of the important features of this maser is its potential for cw operation.

627. Characteristics of $SrF_2:Sm^{2+}$ Optical Masers. P. P. Sorokin, M.J. Stevenson, Jr., R. Lankard, and G. D. Pettit, Bull. Am. Phys. Soc., II, Vol. 7, p. 195, March 1962.

Pulsed optical maser action at temperatures close to 4.2°K is reported. The 6969 A output corresponds to the wavelength of an exceedingly sharp line. The output is characterized by strong relaxation oscillations.

628. Spectroscopy and Optical Maser Action in Strontium Fluoride:Divalent Samarium. P. P. Sorokin, M.J. Stevenson, J. R. Lankard,

and G. D. Pettit, Phys. Rev., Vol. 127, pp. 503-508, July 1962.

Results of spectroscopic and optical maser measurements
in the system are presented. This system differs from
CaF_2:Sm by having a four-orders-of-magnitude longer radiative lifetime. The reasons for the difference in lifetime
values and the effects of this difference on maser kinetics
are presented.

629. Light Scattering from Dielectric Film Laser Mirrors. W. A.
Spechy, Jr., Proc. IEEE, Vol. 51, pp. 615-616, April 1963.

This paper is a report of an investigation into the mirror
scattering and its effect on laser operation.

630. Observations on Oscillation Spikes in Multimode Lasers. H. Statz,
C. Luck, C. Shafer, and M. Ciftan, pp. 342-347 in Advances in
Quantum Electronics, J. R. Singer, ed., Columbia University
Press, New York, 1961.

A theory accounting for oscillation spikes in multimode lasers
is presented.

631. Zeeman Effect in Optical Helium-Neon Maser. H. Statz, R. A.
Paananen, and G. F. Koster, Bull. Am. Phys. Soc., II, Vol. 7,
p. 195, March 1962.

The Zeeman effect due to natural and artificial magnetic
fields has been observed. The maser emission in general
shows two elliptically polarized outputs of different frequencies and sometimes, in addition, a linearly polarized
light having a frequency halfway between the two elliptically
polarized waves.

632. Zeeman Effect in Gaseous He-Ne Optical Maser. H. Statz, R. A.
Paananen, and G. F. Koster, J. Appl. Phys., Vol. 33, pp. 2319-
2321, July 1962.

For a small magnetic field parallel to the maser axis two
circularly polarized components of different frequency are
emitted. Their superposition may be described as linearly
polarized light where the plane of polarization is rotating
at half the difference frequency. A polarizer in the beam
gives rise to amplitude-modulated light.

633. Effect of Spatial Cross Relaxation on the Spectral Output and Spiking Behavior of Solid State Lasers. H. Statz, C. Tang, and G. DeMars, Bull. Am. Phys. Soc., II, Vol. 8, p. 87, January 1963.

Experiments show that conventional solid-state lasers can go into oscillation simultaneously in many modes. Formulas which relate the number of unstable modes to the pump power and various other maser parameters are obtained. The results show that it is difficult to obtain a single-mode operation.

634. Line Widths and Pressure Shifts in Mode Structure of Stimulated Emission from GaAs Junctions. M. J. Stevenson, J. D. Axe, and J. R. Lankard, IBM J. Res. Dev., Vol. 7, pp. 155-156, April 1963.

This communication reports the high-resolution measurements of line width as well as the dependence of the GaAs stimulated emission characteristics on hydrostatic pressure.

635. Spectral Characteristics of Exploding Wires for Optical Maser Excitation. M. J. Stevenson, W. Reuter, N. Braslau, P. P. Sorokin, and A. J. Landon, J. Appl. Phys., Vol. 34, pp. 500-509, March 1963.

As pulsed light sources, exploding wires can be used to provide intense narrow spectral lines. In the visible and UV regions of the spectrum the spectral radiance of air-exploded wires is two to three orders of magnitude greater than that of conventional flash lamps. Vacuum-exploded wires have spectral radiance another factor of eight greater.

636. Exploding Wires as Pumping Sources for Optical Masers. M. J. Stevenson, W. Reuter, P. P. Sorokin, and A. J. Landon, Bull. Am. Phys. Soc., II, Vol. 7, p. 195, March 1962.

A detailed spectrographic analysis of the integrated light output from exploding wires has been carried out under a variety of experimental conditions. The parameters varied were material, energy, wire thickness and length, holder configuration, etc.

637. On the Possibility of Observing Laser Action from the R_2 Line in Ruby. C. M. Stickley AD291738, 9 pp., September 1962.

From a knowledge of the absorption coefficients of the R_1

and R_2 lines it is shown that it is theoretically possible to obtain oscillation from only the R_2 line.

638. Observation of Beats between Transverse Modes in a Ruby Laser. C. M. Stickley, Proc. IEEE, Vol. 51, p. 848, 1963.

The observation of beats between transverse modes of a ruby laser is reported. The following evidence supports this interpretation of the observed phenomenon: 1) the beats appear in time-coincidence with spikes that are emitted by the ruby laser, 2) the probability of observing them increases as the temperature is reduced, and 3) the beat frequencies are within the range predicted by Schawlow and Townes, and by Barone using a different analysis.

639. Radiation Patterns and Axial Modes of Ruby Lasers. C. M. Stickley and R. A. Bradbury, Air Force Cambridge Research Laboratories Rpt. AFCRL-62-386, June 1962.

Improvement of the flatness and parallelism of the ends has been found to have little effect on the beam width and far-field radiation pattern of a ruby laser if the crystal is of poor quality. The axial modes in ruby have been separated optically, and their separation agrees with the basic theory within the limits of experimental error.

640. Time Variation of Axial Frequencies in Ruby Lasers. C. M. Stickley, R. C. White, Jr., and R. A. Bradbury, Air Force Cambridge Research Laboratories, L. G. Hanscom Field, Bedford, Massachusetts, Lasers and Applications Symposium, Ohio State University, November 1962.

Present ruby laser crystals, being only of medium quality, exhibit broad fluorescent line widths, which permit the laser to oscillate at many axial frequencies. The purpose of this paper was to discuss the results of an experiment which aids in describing mode switching in ruby lasers.

641. Observation of Beat Frequencies between Transverse Modes of Ruby Lasers. C. M. Stickley and R. L. Townsend, Polytechnic Institute of Brooklyn Symposia Series, XIII, Optical Masers, April 1963.

This paper reports the observation of beat frequencies between the transverse modes of a ruby laser. They appear in time-coincidence with the spikes that are emitted by the ruby laser. The probability of observing them increases as the temperature is reduced. Beat frequencies are within the range predicted by Schawlow and Townes when the radius of the emitting region is used in the expression for beat frequency instead of the radius of the laser rod.

642. Power Output Characteristics of a Ruby Laser. M. L. Stitch, J. Appl. Phys., Vol. 32, pp. 1994-1999, October 1961.

The theoretical power output of a ruby laser is examined under certain idealized operating conditions and the two principal regions of operation, the regions of strong oscillation and saturation, are described. The efficiency of operation is examined under two limiting conditions.

643. Microwave Interaction with Matter. M. L. Stitch, Proc. IRE, Vol. 50, pp. 1225-1231, May 1962.

The interaction between microwaves and matter is illustrated by two classes of phenomena: those in which the microwave field loses energy and those in which it gains energy. The latter is explained by the technique of molecular beams. The extension of the maser principle into the infrared and optical regions is discussed.

644. Repetitive Hair-Trigger Mode of Optical Maser Operation. M. L. Stitch, E. J. Woodbury, and J. H. Morse, Proc. IRE (Correspondence), Vol. 49, pp. 1570-1571, October 1961.

A method is proposed which would permit high-repetition-rate pulse operation of a three-level optical maser in a periodic, predictable, and controllable manner without the use of shutters.

645. Stimulation Versus Emission in Ruby Optical Maser. M. L. Stitch, E. J. Woodbury, and J. H. Morse, pp. 83-84 in Advances in Quantum Electronics, J. R. Singer, ed., Columbia University Press, New York, 1961.

Emission from ruby optical masers is examined under varying excitation power at fixed excitation energy. Data from several full-aperture partially transparent coated rubies differed markedly from that obtained from an opaque coated ruby with a small aperture in the center. A hypothesis to explain the results is advanced.

139

646. New Method for Achieving Mode Discrimination in Solid-State Lasers. T. L. Stocker and M. Birnbaum, Bull. Am. Phys. Soc., II, Vol. 8, p. 443, 1963.

> Ruby laser structures formed by butting together two ruby laser rods with plane-parallel ends have been shown to possess mode selection properties. The interference between the two rods acts like a partially transparent reflector.

647. Interferometric Studies of Ruby Maser Emission. B. P. Stoicheff and G. R. Hanes, J. Opt. Soc. Am., Vol. 52, p. 595, May 1962.

> The wavelength and line width characteristics of the optical emission from a ruby maser have been studied at various temperatures and power levels.

648. Spectroscopy for Solid State Optical Masers. S. Sugano, Appl. Optics, Vol. 1, pp. 295-301, May 1962.

> A brief account is given of the spectroscopy of excitation processes in semiconductors, rare earth ions, transition metal ions, and actinide group ions in ionic crystals, from the standpoint of applying solid-state optical spectra to optical masers.

649. Absorption Spectra of Cr^{3+} in Al_2O_3. Part A. Theoretical Studies of the Absorption Bands and Lines. S. Sugano and Y. Tanabe, J. Phys. Soc. (Japan), Vol. 13, pp. 880-899, August 1958.

> In the framework of the crystalline field theory the excited states of Cr^{3+} in Al_2O_3 and the optical transitions to these states are studied, taking into account the effect of trigonal field and spin-orbit interaction.

650. Pumping Power Considerations in an Optical Maser. O. Svelto, ML Report No. 902, AD275479, 29 pp., Stanford University Microwave Laboratory, Stanford, Calif., 1962; Appl. Optics, Vol. 1, pp. 745-751, 1962.

> A thermodynamical method is used to calculate power transfer from the flash tube to a laser rod. A general expression is obtained for calculations of the power required from the flash-tube for a given power absorption by the rod.

651. Pumping Power Considerations in an Optical Maser. O. Svelto,

Appl. Optics,Supplement 1, pp. 107-113, 1962.

> A thermodynamic method is used to calculate power transfer from the flash tube to a laser rod. A general expression is obtained for calculations of the power required from the flash tube for a given power absorption by the rod.

652. Microwave Maser Action in Ruby at 78°K by Laser Pumping. A. Szabo, Polytechnic Institute of Brooklyn Symposia Series, XIII, Optical Masers, April 1963.

> Maser amplification and oscillation at X-band have been observed in ruby using a ruby laser as a pump. The laser frequencies were matched to the required optical transitions in the microwave ruby by thermal tuning of the latter ruby. Measurements of the thermal tuning rate of the R_1 line in ruby using the paramagnetic resonance absorption of the ground-state levels for detection will be described as well as the conditions under which microwave maser action was obtained.

653. Side Emission from Ruby Laser Rods. A. Szabo and F. R. Lipsett, Proc. IRE, Vol. 50, p. 1690, July 1962.

> The measurements are compared to similar measurements in $CaF_2 : Sm$.

654. On Diffraction Losses in Laser Interferometers. C. L. Tang, Appl. Optics, Vol. 1, pp. 768-770, November 1962; Raytheon Tech. Memo. T320.

> Diffraction losses in the infinite-strip and circular-disc types of laser interferometers are determined analytically using a variational method. Numerical results obtained are shown to be in close agreement with the computer solutions of Fox and Li.

655. Relative Probabilities for the Xenon Laser Transitions. C. L. Tang, Proc. IEEE (Correspondence), Vol. 51, pp. 219-220, January 1963.

> Transition probabilities are calculated for xenon.

656. Optical Harmonic Generation in Calcite. R. W. Terhune, P. D.

Maker, and C. M. Savage, Phys. Rev. Lett., Vol. 8, pp. 404-405, May 1962.

Optical second harmonic generation has been observed as a function of an applied dc electric field in crystals of calcite, which possess a center of inversion. Of particular significance is that a small amount of doubling was observed to occur with no dc electric field applied.

657. Observation of Saturation Effects in Optical Harmonic Generation. R. W. Terhune, P. D. Maker, and C. M. Savage, Appl. Phys. Lett., Vol. 2, pp. 54-55, February 1963.

A large increase in optical harmonic generation is observed using a giant-pulse ruby laser. It appears that harmonic generation is an excellent technique for that harmonic generation is an excellent technique for obtaining laser beams in the ultraviolet region.

658. Optical Pumping in Crystals. H. H. Theissing, P. J. Caplan, F. A. Dieter, and N. Rabbiner, Phys. Rev. Lett., Vol. 3, pp. 460-462, November 1960.

An estimate is made of the expected population change in the ground state of a crystal due to optical pumping.

659. A Method for Evaluating Laser Potentialities of Crystals. H. H. Theissing, P. J. Caplan, T. Ewanizky, and G. deLhery, Appl. Optics, Vol. 2, pp. 291-298, March 1963.

Optical measurements of a figure of merit with reference to a particular pumping device are described which can be performed on slabs without laser end faces. Such measurements can be carried out by saturation with white light in which the fluorescence wavelength under study has been removed.

661. Fluorescence in CdS and Its Possible Use for an Optical Maser. D. G. Thomas and J. J. Hopfield, J. Appl. Phys., Vol. 33, pp. 3243-3249, November 1962.

This paper discusses how the fluorescence from semiconductors might be used in constructing an optical maser. Attention is given to the sharp line emission which occurs at low temperatures in CdS and which arises from excitons bound to impurities.

662. Paramagnetic Resonance of the Shallow Acceptors Zn and Cd in GaAs. R. S. Title, IBM J. Res. Dev., Vol. 7, pp. 68-69, January 1963.

The emission observed in GaAs diode injection lasers is believed to be an electronic transition involving acceptor levels. In this letter the paramagnetic resonance absorptions of the neutral acceptors Zn and Cd in GaAs are reported.

663. Biological Effects of Concentrated Laser Beams. V. T. Tomberg, Polytechnic Institute of Brooklyn Symposia Series, XIII, Optical Masers, April 1963.

The biological and medical applications of a coherent light source such as that of a laser are reviewed. A coordinated light source is capable of building up a strong electrical field of molecular size and is therefore capable of inducing biological effects of an electrical nature independent of the thermal effect as used in coagulation.

664. Focused Side Pumping of Laser Crystal. K. Tomiyasu, Proc. IRE, Vol. 50, pp. 2488-2489, December 1962.

An elliptical configuration is analyzed as an example of focused side pumping.

665. Filamentary Standing-Wave Patterns in a Solid-State Maser. L. Tonks, J. Appl. Phys., Vol. 33, pp. 1980-1986, June 1962.

The filamentary maser action in a ruby is explained as caused by minute randomly distributed variations in index of refraction. Relations are derived for the density of filaments and their size in terms of a Fourier resolution of the inhomogeneities. These are so small that it seemed thermal motion might play a part, but this is shown to be orders of magnitude too small at room temperature.

666. Quantum Electronics. C. H. Townes, ed., Columbia University Press, New York, 606 pp., 1960.

This volume represents papers and discussion at the Conference on Quantum Electronics — Resonance Phenomena held at Shawanga Lodge, High View, New York, in September 1959.

667. Optical and Infrared Masers. C. H. Townes, J. Opt. Soc. Am., Vol. 51, p. 471, April 1961.

Theoretical and experimental work in the field of optical and infrared masers is reviewed.

668. Some Applications of Optical and Infrared Masers. C. H. Townes, pp. 3-11 in Advances in Quantum Electronics, J. R. Singer, ed., Columbia University Press, New York, 1961.

Five experiments are considered: interplanetary communications, frequently multiplication, high-resolution spectroscopy in the microwave-near infrared region, high-resolution Raman spectroscopy, and the precise comparison of lengths and frequencies.

669. Theory of Relaxation Spikes in Two-Level Laser Amplifiers. H. A. Trenchard, Air Arm Division, Westinghouse Electric Corporation, Baltimore, Maryland, Third International Symposium on Quantum Electronics, Paris, France, February 1963.

This paper considers the interaction of an electromagnetic wave which is propagated through an initially inverted two-level system. Numerical solutions for the output have been obtained for ruby and the results show that relaxation spikes occur in the output beam of the laser.

670. Infrared Masers Using Rare Earth Ions. G. J. Troup, pp. 85-90 in Advances in Quantum Electronics, J. R. Singer, ed., Columbia University Press, New York, 1961.

This paper discusses possible active media for infrared masers and ignores the interaction space. Only the solid-state maser is considered since it is known to be the most efficient in terms of space utilized.

671. Induced Gamma-Ray Emission. V. Valli and W. Valli, Proc. IEEE, Vol. 51, pp. 182-185, January 1963.

The extension of optical maser techniques to the gamma-ray region is considered. It is shown that under certain conditions induced gamma rays can be produced. The condition of criticality rather than that of oscillation is used because a gamma-ray maser does not have a resonant structure. The main observable effects are the shortening of lifetimes of some gamma excitations and the appearance of two or more coherent gamma quanta. Principles are applied to a specific example.

672. Self-Adsorption and Trapping of Sharp-Line Resonance Radiation in Ruby. F. Varsanyi, D. L. Wood, and A. L. Schawlow, Phys. Rev. Lett., Vol. 3, pp. 554-555, December 1959.

The details of the sharp-line fluorescence of ruby are examined. It has been possible to observe the trapping of resonance radiation in a solid for the first time in 0.05% ruby.

673. Excitation of Fluorescence with Monochromatic Light in Rare-Earth Crystals. F. Varsanyi, Bell Telephone Laboratories, Murray Hill, N. J., Third International Symposium on Quantum Electronics, Paris, France, February 1963.

Changes in the fluorescence spectra of rare earth crystals were studied as a function of the excitation frequency. Most attention was paid to $LaCl_3$ containing a few percent of other rare earths. In some cases, the excitation spectrum and the absorption spectrum above a fluorescing level are markedly different. Apparently the efficiency of the energy transfer changes, but it does not fall off uniformly as the frequency difference between fluorescing level and the excitation radiation increases.

674. Crystal Field Effects in Solid State Sources of Laser Action. R. C. Vickery, Semi-Elements, Saxonburg, Penna., Third International Symposium on Quantum Electronics, Paris, France, February 1963.

The effect of crystalline fields upon absorption spectra of rare earth ions in solid state matrices has been studied. The variations developed have been examined on the basis of crystal configuration and data thus obtained correlated with laser activity parameters. Proposals are made for idealized structures permitting higher operational efficiencies. Experimental data following such proposals are presented.

675. The Skelaser Lamp. A Unitary Design for the Emission of Stimulated Radiation. R. C. Vickery and J. V. Fisher, Semi-Elements, Saxonburg, Pennsylvania, Third International Symposium on Quantum Electronics, Paris, France, February 1963.

> Laser action is developed in an integrated unit of novel design which utilized the emitting crystal as the material from which the exciting light source is fabricated. The design eliminates the necessity for external pumping sources such as helical flash tubes and improves the efficiency of operation as well as utilizing energy at wavelengths normally screened off by the flash tube envelope.

676. Les Maser Optique. J. C. Viénot, Rev. Optique (France), Vol. 40, pp. 4-22, January 1961 (in French).

> The properties of the induced emission process lead to the consideration of ensembles of molecular systems satisfying resonance conditions and capable of producing a monochromatic and coherent radiation. A brief account is given of the fundamental principles and some of the specific methods for the optical domain are described, as well as recent investigations. An approach dealing with chemical reaction processes is mentioned.

677. Determination of the Width of the R_1 Line (6943 A) Emitted by a Ruby Laser. J. C. Viénot, N. Aebischer, and J. Bulabois, Compt. rend. acad. sci. (France), Vol. 24, pp. 1596-1598, February 1962 (in French).

> The fringes formed at infinity by a Fabry-Perot interferometer illuminated with the light emitted by a ruby laser indicate that the spectral distribution of the emission is less than 0.004 cm^{-1}. An uncertainty caused by a granular structure due to discontinuities in the phase of light emitted from different parts of the face of the crystal is included.

678. Theory of Laser Regeneration Switching. A. A. Vuylsteke, Bull. Am. Phys. Soc., II, Vol. 7, p. 553, November 1962.

> The theory of laser regeneration switching is discussed in terms of two coupled rate equations. The material cavity parameters are evaluated for ruby. Approximate solutions indicate three modes of operation other than the normal mode.

679. Cavity Modes in an Optical Maser. W. G. Wagner and G. Birnbaum, Proc. IRE, Vol. 49, p. 625-626, March 1961.

> The authors find that in the ideal case above a certain threshold of pumping power a system of quantum oscillators in a multimode cavity breaks into oscillation in the lowest-order mode. The model is based on a three-level system.

680. Theory of Quantum Oscillators in a Multimode Resonator. W. G. Wagner and G. Birnbaum, J. Opt. Soc. Am., Vol. 51, p. 472, April 1961.

> The spectrum of power radiated by a solid-state optical maser in steady-state operation has been obtained by considering each atomic system to be a source of randomly fluctuating dipole moment which drives every mode of the cavity. The nonlinear behavior of the collection of atomic systems has been treated in such a way that a detailed examination of the distribution of power in the various modes is possible.

681. Theory of Quantum Oscillators in a Multimode Cavity; Solid State Optical Maser in Steady-State Operation. W. G. Wagner and G. Birnbaum, J. Appl. Phys., Vol. 32, pp. 1185-1194, July 1961.

> The spectrum of power radiated by a solid-state cw laser is obtained by considering each atomic system to be a source of randomly fluctuating dipole moment which drives every mode of the cavity. The nonlinear behavior of the collection of atomic systems is treated in such a way that a detailed examination of the distribution of power in the various modes is possible.

682. A Steady-State Theory of the Optical Maser. W. G. Wagner and G. Birnbaum, pp. 328-333 in Advances in Quantum Electronics, J. R. Singer, ed., Columbia University Press, New York, 1961.

> The optical maser oscillator and amplifier are discussed. Some aspects of previous work are amplified and the formalism developed previously for the case of the optical maser amplifier is extended.

683. Laser Operation without Spikes in a Ruby Ring. P. Walsh and G. Kemeny, J. Appl. Phys., Vol. 34, pp. 956-957, April 1963.

Pulsed laser operation has been produced by complete internal reflection in a ruby ring at room temperature. The linearized equations of Statz and de Mars give a good description of the steady-state laser operation. The lifetime of the excited Cr ions is found to be one microsecond. Short-lived transient oscillations are observed in the output.

684. Laser Operation in a Ruby King. P. Walsh and G. Kemeny, Bull. Am. Phys. Soc., II, Vol. 7, pp. 397-398, June 1962.

Pulsed laser operation by internal reflection in a ruby ring at room temperature is described. No spiking is observed.

685. Radiation Patterns of Confocal He-Ne Laser. W. R. Watson and T. G. Polanyi, J. Appl. Phys., Vol. 34, pp. 708-709, March 1963.

Experimental results indicate that the radiation patterns result from simultaneous oscillation in many modes, the rectangular arrays in lasers having Brewster angle windows are not produced by the windows, the higher degree of circular symmetry exhibited in patterns from external lasers having perpendicular windows in the cavity results from a strong selection against the more common rectangular modes, and confocal lasers give more complicated patterns because they support a large number of modes.

686. Masers. J. Weber, Rev. Mod. Phys., Vol. 31, pp. 681-710, July 1959.

The principles of maser-type amplification are presented. Reference is made to the possibility of obtaining maser action in the optical region.

687. Optics for the Optical Maser. P. E. Weber, J. Opt. Soc. Am., Vol. 52, p. 602, May 1962.

The topics discussed include the design of thick lenses with minimum spherical aberration and coma; effects of high-power density laser beams on cemented optics and concentrating lens systems; optical systems for imaging high-intensity light patterns into small target areas; methods of obtaining spatial and time coherence information; techniques for measuring the end parallelism of laser rods and the mirrors of gaseous lasers.

688. Spatial Distribution of Light Across the End of a Ruby Laser. D. Weisman, Appl. Optics, Vol. 1, pp. 672-673, September 1962.

> A criticism of current articles on the spatial distribution of light across the end of a laser is presented.

689. Eight-Inch Ruby Amplifier. J. L. Wentz, Proc. IRE, Vol. 50, p. 1528, June 1962.

> A twin elliptical cavity is used to excite a ruby cylinder eight inches long.

690. Pulsed Laser Performance Prediction. Joseph H. Wenzel, General Electric Company, Ithaca, New York, Lasers and Applications Symposium, Ohio State University, November 1962.

> This paper presents a set of equations for predicting the pulsed output energy of a laser which should be useful in the design of efficient high-power sources. A mathematical model of a laser based upon volume average absorbed pump power and spontaneous and stimulated emission is described. Computational examples of high-power configurations are also presented, and a comparison between predicted and typical existing laser performance is made.

691. Prediction of Q-Switch Laser Energy Output. J. Wenzel, Polytechnic Institute of Brooklyn Symposia Series, XIII, Optical Masers, April 1963.

> This paper derives a set of equations for predicting the energy output of a Q-switch or controlled reflectance laser. A mathematical model of the stored energy laser oscillator, based upon volume-average absorbed pump power and the limiting effects of amplified spontaneous emission, is described. Computational examples of the predicted performance of several Q-switched lasers are presented, and a comparison is made of predicted and actual performance.

692. Investigations of Optics for a Radiation Weapon System, Vol. III. Westinghouse Electric Corporation, AD291466, 256 pp., August 1962.

> Included are articles on laser materials, laser pumping,

and field patterns of focused continuous apertures and aperture arrays.

693. Simultaneous Gas Maser Action in the Visible and Infrared. A. D. White and J. D. Rigden, Proc. IRE, Vol. 50, pp. 2366-2367, November 1962.

    A nearly confocal He-Ne maser with external mirrors has been made to oscillate simultaneously at 6328 A and at wavelengths in the infrared.

694. Output Power of the 6328 A Gas Maser. A. D. White, E. I. Gordon, and J. D. Rigden, Appl. Phys. Lett., Vol. 2, pp. 91-93, March 1963.

    Measurements of the output power of a dc excited He-Ne gas maser were made as a function of active plasma length and discharge current.

695. Continuous Gas Maser Operation in the Visible. A. D. White and J. D. Rigden, Proc. IRE, Vol. 50, p. 1697, July 1962.

    The results of spectral studies of the helium-neon gas discharge plasma are given.

696. Optical Detection of Paramagnetic Resonance Saturation in Ruby. I. Wieder, Phys. Rev. Lett., Vol. 3, pp. 468-470, November 1959.

    R radiation from a ruby (at liquid nitrogen temperature) of intensity 0.1 W/R_{12} line was collected, polarized, and focused on a second ruby at liquid helium temperature within the cavity of a microwave spectrometer.

697. Some Microwave-Optical Experiments in Ruby. I. Wieder, pp. 105-109 in Quantum Electronics. C. H. Townes, ed., Columbia University Press, New York, 1960.

    The effects of optical pumping in ruby are investigated.

698. Stimulated Optical Emission from Exchange-Coupled Ions of Trivalent Chromium in Aluminum Oxide. I. Wieder and L. R. Sarles, Phys. Rev. Lett., Vol. 6, pp. 95-96, February 1961.

    The observation of stimulated emission at wavelengths of

7010 and 7040 A from transitions in red ruby which arise from exchange coupling between neighboring chromium ions is reported.

699. Relaxation Times between Excited States of Ruby. I. Wieder and L. R. Sarles, J. Opt. Soc. Am., Vol. 51, p. 473, April 1961.

To investigate the rate of nonradiative transfer of atoms from upper to lower excited states, the relative intensities of R_1 and R_2 fluorescence have been monitored following a pulse of exciting radiation. The observed intensity ration of fluorescent R_1 and R_2 light did not deviate from the Boltzmann distribution thus placing a limit on the thermal relaxation time between the two levels.

700. Quantum-Mechanical Effects in Stimulated Optical Emission. R. C. Williams, Appl. Optics,Supplement 1, pp. 63-66, 1962; Phys. Rev., Vol. 126, pp. 1011-1014, May 1962.

Two distinct photon-emission processes take place within a three-level system when stimulated optical emission occurs. A two-photon transition consists of the emission of a stimulated photon followed by the absorption of a coherent pump photon. In narrow-pump-level three-level systems two-photon transitions are dominant at high pump powers while the maser line is split into two lines due to the modulation of the wave function at an angular frequency determined by the rate of pumping.

701. Liquid Laser Research. M. W. Windsor, AD291592, 58 pp., October 1962.

Results highly suggestive of laser action have been obtained for benzophenone and a chelate of terbium in rigid organic glasses at 77°K. Experimental techniques are described.

702. Effects of Elevated Temperatures on the Fluorescence and Optical Maser Action of Ruby. J. P. Wittke, J. Appl. Phys., Vol. 33, pp. 2333-2335, July 1962.

Measurements are made of the wavelengths, linewidth, and fluorescent efficiencies of the R_1 and R_2 lines in ruby in the temperature range 300 to 500°K. The results are used to explain the degeneration of ruby optical maser performance at elevated temperatures.

703. Uranium-Doped Calcium Fluoride as a Laser Material. J. P. Wittke, Z. J. Kiss, R. C. Duncan, Jr., and J. J. McCormick, Proc. IEEE, Vol. 51, pp. 56-62, January 1963.

Spectroscopic studies have shown that $U:CaF_2$ crystals exhibit two types of U^{3+} sites. The first of these sites gives rise to a maser transition at 2.51 microns, the second to transitions at 2.61 and 2.57 microns.

704. Is a Complete Determination of the Energy Spectrum of Light Possible from Measurements of the Degree of Coherence? E. Wolf, Proc. Phys. Soc., Vol. 80, pp. 1269-1272, December 1962.

It is shown that the analytic properties of the temporal complex degree of coherence in the complex time plane imposes relationship between its magnitude and argument on the real time axis. This relationship involves the location of the zeros of the degree of coherence in the lower half of the complex plane. It is suggested that the analytic continuation of the temporal degree of coherence of many spectral distributions has no zeros at all in this half plane. Spectral profiles of such distributions could be uniquely determined by one such measurement.

705. Basic Concepts of Optical Coherence Theory. E. Wolf, Polytechnic Institute of Brooklyn Symposia Series, XIII, Optical Masers, April 1963.

An account is given of the basic concepts of optical coherence theory and of the extensions currently proposed by a number of authors.

706. Optical Maser Action in an Eu^{3+}-Containing Organic. N. E. Wolff and P. J. Pressley, Appl. Phys. Lett., Vol. 2, pp. 152-153, April 1963.

A system containing a solid solution of europium tris(4,4,4-trifluro-1-(2-thienyl)-1,3-butanedione) in polymethylmethacrylate has shown optical maser action. Probable energy levels are included.

707. Amplitude and Frequency Control in Solid State Optical Masers.

D. L. Wood, National Physical Laboratory, Teddington, Middlesex, U. K., Third International Symposium on Quantum Electronics, Paris, France, February 1963.

The control of the spiking phenomenon in solid state optical masers presumably can be achieved by the use of a nonlinear optical element whose properties include an increase in attenuation with increasing transmitted light flux. Several practical systems are considered.

708. Absorption and Fluorescence of Divalent Samarium in CaF_2, SrF_2, and BaF_2. D. L. Wood and W. Kaiser, Phys. Rev., Vol. 126, pp. 2079-2080, June 1962.

The absorption and fluorescence of the divalent samarium ion have been studied for the three host lattices. Measurements of intensity, line width, quantum efficiency, and Zeeman splitting are reported. Energy levels belonging to the 4f shell have been identified, and a preliminary analysis of the 4f to 5d transition is presented.

709. Ruby Laser Operation in the Near Infrared. E. J. Woodbury and W. K. Ng, Proc. IRE, Vol. 50, p. 2367, November 1962.

A stimulated emission at approximately 7670 A has been observed accompanying the usual 6943 A emission.

710. Lifetimes of Nd^{3+}-Doped Silicate Laser Glasses. R. F. Woodcock, J. Opt. Soc. Am., Vol. 53, p. 523, 1963.

For silicate glasses containing various amounts of trivalent Nd the fluorescent lifetimes of some glasses have been observed to increase and others to decrease on lowering the temperature.

711. Fabrication of High-Efficiency Laser Cavities. B. W. Woodward and G. J. Wolga, Rev. Sci. Instr., Vol. 33, pp. 1463-1465, December 1962.

Fabrication techniques for any cylinder of revolution are presented.

712. Transparent Materials for the Far Infrared (50-1000 micron); Application to Optical Masers. B. Wyncke and A. Hadni, Faculté des

Sciences, Nancy, France, Third International Symposium on Quantum Electronics, Paris, France, February 1963.

> For a thickness of a few centimeters there is not any material known to be transparent in the far infrared. In liquid helium it is found that the absorption on the low-frequency wing of the main absorption band for some crystals disappears almost completely. This result seems to show that the low-frequency absorption may be ascribed to transitions between acoustical and optical branches and that a preliminary thermal excitation is necessary to give a sufficient population to the lower level of the transition.

713. High-Speed Photography Using a Ruby Optical Maser. T. Yajima, F. Shimizu, and K. Shimoda, Appl. Optics, Supplement 1, pp. 137-138, December 1962; Appl. Optics, Vol. 1, pp. 770-771, 1962.

> An example of high-speed photography including the experimental setup is given.

714. Cylindrical Mode of Oscillation in a Ruby Optical Maser. T. Yajima, F. Shimizu, and K. Shimoda, Polytechnic Institute of Brooklyn Symposia Series, XIII, Optical Masers, April 1963.

> A particular type of mode in optical maser oscillation has been obtained with a ruby rod having parallel-plate resonator with a central hole in one of the reflector coatings. The observed far-field pattern of single annular shape, accompanied by fine radial fringes, was interpreted on the basis of Fraunhofer diffraction theory.

715. Continuous Operation of a $CaF_2:Dy^{2+}$ Optical Maser. A Yariv, Proc. IRE, Vol. 50, pp. 1699-1700, July 1962.

> Continuous operation of the maser is reported. Information on the preparation of crystals, threshold energies, concentration of impurity ions, and power output is included.

716. The Laser. A. Yariv, and J. P. Gordon, Proc. IEEE, Vol. 51, pp. 4-29, January 1963.

> A review of the field of optical masers summarizing both theory and practice is presented.

717. Power Output and Optimum Coupling in Continuous Solid State

Lasers. A. Yariv, Bell Telephone Laboratories, Murray Hill, N. J., Third International Symposium on Quantum Electronics, Paris, France, February 1963.

A theoretical derivation is presented which results in expressions for the power output of three- and four-level cw masers. It is shown that these results can be expressed in terms of fluorescence power at threshold and the amount by which threshold is exceeded. Equations for optimum coupling is obtained.

718. Dielectric-Waveguide Mode of Light Propagation in p-n Junctions. A. Yariv and R. C. Leite, Appl. Phys. Lett., Vol. 2, pp. 55-57, February 1963.

The presence of free-charge carriers in p and n regions and their absence from the depletion layer leads to a discontinuity in the real part of the dielectric constant at the junctions boundaries which can give rise to energy confinement via the dielectric waveguide effect.

719. Optical Maser Emission from Trivalent Praseodymium in Calcium Tungstate. A Yariv, S. P. Porto, and K. Nassau, J. Appl. Phys. Vol. 33, pp. 2519-2521, August 1962.

Coherent emission at 1.047 micron from trivalent praseodymium in calcium tungstate was observed. This emission coincides with strong infrared fluorescence at the same wavelength and was found to be stimulated by absorption of blue light by the $3p_0$, $3p_1$, and $3p_2$ bands. The oscillation threshold was the same at 4.2, 20, and 78°K. The lifetime of the metastable state $^1G_4$ is 50×10^{-2} sec. A new technique used to measure the lifetime is described.

720. Anti-Stokes Fluorescence as a Cooling Process. S. Yatsiv, Advances in Quantum Electronics, Columbia University Press, New York, 1961.

Cases favorable to anti-Stokes cooling are pointed out. Limitations of principles for the cooling process are continued and expressions for the cooling rate for a specific example are formulated. The cooling rate is found proportional to intensity of the incident "monochromatic" radiation. The potentialities of utilizing optical means for cooling purposes are strongly dependent on the availability of intense monochromatic sources.

721. Continuous Glass Laser. C. G. Young, Appl. Phys. Lett., Vol. 2, pp. 151-152, April 1963.

Experimental methods for obtaining continuous laser action in barium crown glass are described.

722. Threshold–Pumping Characteristics of the Neodymium Glass Laser. C. G. Young, J. Opt. Soc. Am., Vol. 52, p. 1318, 1962.

A study of factors important to pumping a glass laser has been made. Xenon and neon flash lamps were used with a range of filters to isolate the most efficient pumping regions. A variety of exploding wires was also used in an attempt to obtain greater variety in spectral output.

723. Paramagnetic Electron Resonance. P. Yung, Rev. E. (Belgium), Vol. 3, pp. 271-277, 1961 (in French).

The fundamentals of paramagnetic resonance and their applications in masers and lasers are discussed.

724. Biomedical Experimentation with Optical Masers. M. M. Zaret, H. Ripps, I. Siegel, and G. M. Breinin, J. Opt. Soc. Am., Vol. 52, p. 607, May 1962.

Ocular abiotic effects of a pulsed ruby maser were investigated. The retina and iris of the rabbit eye exhibited instantaneous thermal injury following exposure.

725. Characteristics of the Ruby Laser. C. Zarowin and R. L. Martin, J. Opt. Soc., Vol. 51, p. 476-477, April 1961.

A number of definite cavity modes have been observed in the beam of the ruby laser. Off-axis modes appear to contribute to the observed beam width of 15 min of arc. The analysis of these modes as well as the conditions under which they are observed is discussed.

726. The Use of Resonant Cavity Spectroscopy to Study the Populations in the He-Ne Systems. C. B. Zarowin, M. Schiff, and G. R. White, Polytechnic Institute of Brooklyn Symposia Series, XIII, Optical Masers, April 1963.

A. Kastler has described the properties of a Fabry-Perot cavity containing fluorescent atoms. The authors have extended his observations and developed a technique which provides a measure of the relative populations of an ensemble of atoms within the Fabry-Perot cavity. This technique (Resonant Cavity Spectroscopy) has been applied to the study of the $3S_2$ to $2P_{10}$ transition at 5433 A (Paschen notation) of neon in the He-Ne system.

727. Dynamics of the Fields in a Laser Oscillator. C. B. Zarowin and C. C. Wang, Sperry Gyroscope Co., Great Neck, N. Y., Third International Symposium on Quantum Electronics, Paris, France, February 1963.

The solutions to the rate equations are introduced into the continuity equation for a forward- and a backward-traveling wave in a differential volume of active medium. Computer solutions of the rate equations and their effect on the transmission line equation solutions are presented.

728. Experimental Investigation of Fabry-Perot Interferometers. R. W. Zimmerer, Proc. IEEE, Vol. 51, pp. 475-476, March 1963.

Preliminary measurements of the microwave performance of Fabry-Perot interferometers with spherical mirrors are presented and compared with theory.

729. Quarterly Progress Report No. 64 (period ending Nov. 30, 1961). H. J. Zimmerman, G. G. Harvey, and W. B. Davenport, Jr., MIT Research Laboratory of Electronics, Cambridge N62-12578, 380 pp., 1962.

Research objectives are reviewed.

730. Quarterly Progress Report No. 66 (period ending May 31, 1962). H. J. Zimmerman, G. G. Harvey, and W. B. Davenport, Jr., MIT Research Laboratory of Electronics, N62-14753, 460 pp., 1962.

The status of research on optical and infrared masers is reviewed.

731. Rate Analysis of Multi-Step Excitation in Mercury Vapor. R. Zito, Westinghouse Electric Corporation Research Laboratory, Pittsburgh, Penna., Third International Symposium on Quantum Elec-

tronics, Paris, France, February 1963.

In the course of studying a multiple-step excitation process in mercury vapor for the possible application to three-dimensional display of data, an analysis of the fluorescence mechanism was made. A dual optical excitation process is employed by which mercury vapor is caused to radiate at a localized region in space. The radiation is in the ultraviolet and visible portions of the spectrum.

BIBLIOGRAPHIES

732. Masers and Lasers. E. H. Hall, Comp. Armed Services Technical Information Agency, Arlington, Va., AD271100, 1962. An ASTIA report bibliography.

733. Masers and Lasers: a Bibliography. J. F. Price and A. K. Dunlap, Space Technology Laboratories, Inc., Redondo Beach, Calif., Research bibliography No. 41; 9990-6052-KU-000; AD274843, 161 pp., 1962.

734. Maser and Laser and Iraser: a Bibliography. HRD-Singer, Inc., State College,Pa., Report No. B-1; AD275440, 28 pp., 1962.

735. Soviet Gas Laser Research: a Review of Open Literature. AID Report 62-100.

736. Bibliography on Lasers and Masers. Armed Services Technical Information Agency, Arlington, Va., ARB-9156, 105 pp., 1961.

737. A Bibliography of the Open Literature on Lasers. E. V. Ashburn, Technical Note 4034-2, U. S. Naval Ordnance Test Station, China Lake, California, October 1962.

738. Bibliography on Lasers 1958-1962. Becker and Warren, Clinton Courier-News, Clinton, Tennessee, 1962. 514 laser articles are included. Published by Becker and Warren, Oak Ridge, Tenn.

739. Optical Communications: a Bibliographic Survey of Possible Space and Terrestrial Applications of the Laser and Maser. J. B. Goldman, Report AD275591, OTS, Dept. of Commerce, Washington 25, D. C., 54 pp., March 1962.

740. Masers and Lasers. C. A. Hogg and L. C. Sucsy, Maser/Laser Assoc., Cambridge 40, Mass.

741. Masers and Lasers: a Bibliography. OTS, Dept. of Commerce, Washington 25, D. C., SB-488. A bibliography of U. S. Govt. Re -

search Reports and other material.

742. A Laser Bibliography. K. J. Spencer, TIL3, U. K. Ministry of Aviation, Central Library, London, England, December 1958 - May 1962.

743. Masers, a Literature Search. H. D. Raleigh, Comp. U. S. Atomic Energy Commission, Division of Technical Information Extension, TID 3566-1961.

AUTHOR INDEX

161

Blumenthal, R. H. - 61
Boatright, A. - 62
Boeroff, D. L. - 497
Bond, W. - 63, 64, 115, 214, 216, 298
Boot, H. A. - 65
Borie, J. C. - 66
Bostik, H. A. - 67
Bowen, D. E. - 337
Bowness, C. - 68, 69
Boyd, G. D. - 70, 71, 72, 73, 304, 305, 306, 307, 308
Boyle, W. S. - 348, 480
Bozman, W. R. - 127
Bracewell, R. N. - 74
Bradbury, R. A. - 639, 640
Brand, F. A. - 75, 289
Brangaccio, D. J. - 76, 547
Braslau, N. - 608, 609, 635
Braunbeck, J. - 77
Breinin, G. M. - 724
Brock, E. G. - 78
Bronco, C. J. - 282
Brookman, J. W. - 409
Browne, P. F. - 79
Bruma, M. S. - 80
Buck, A. - 81
Buddenhagen, D. A. - 82
Buhrer, C. F. - 60, 83, 84
Bulabois, J. - 677
Burch, J. M. - 85
Burgess, J. H. - 549
Burgess, J. Q. - 86
Burns, G. - 87, 88, 89, 90, 91, 92, 382, 475, 476, 477
Butayeva, F. A. - 93, 183
Byerly, E. H. - 94

C

Calviello, J. A. - 95
Caplan, P. J. - 658, 659
Carlson, R. D. - 244
Carnahan, C. W. - 96, 97
Carpenter, D. R. - 50
Chang, W. S. C. - 86, 557

Chen, D. - 98, 538
Chizikovu, Z. A. - 206
Christie, R. H. - 99
Church, C. H. - 100, 101, 102, 103
Ciftan, M. - 104, 105, 106, 375, 376
Clark, G. L. - 107, 542
Clunie, D. M. - 65
Cohen, B. G. - 64
Colgate, S. A. - 108, 109
Colling, R. J. - 221
Collins, R. J. - 71, 110, 111, 112, 113, 114, 115, 481, 482
Collins, S. A. - 116, 117, 118
Condell, W. J. - 119, 120
Congleton, R. S. - 121, 620
Connes, P. - 287
Conwell, E. - 83
Cook, J. C. - 122, 123, 124, 125
Corcoran, V. J. - 126
Corliss, C. H. - 127
Cox, G. C. - 31
Craig, D. P. - 128
Crevier, R. - 615
Crowe, J. W. - 129
Cullen, A. L. - 130, 131, 132
Culshaw, W. - 133, 134
Cummins, H. Z. - 6, 135, 136
Curcio, J. A. - 366
Cutler, S. - 584

D

Daly, R. T. - 137, 138
Damon, E. K. - 139, 140, 141
Davenport, W. B., Jr. - 729, 730
Davis, D. T., Jr. - 410, 411
Davis, L. W. - 142
Davison, W. F. - 143
Dayhoff, E. S. - 144, 145, 146, 147, 148
Debye, P. P. - 376
Dekinder, R. E. - 338
DeLhery, G. - 659
DeMaria, A. J. - 149, 150, 151
DeMars, G. - 633

162

296, 297, 298, 299, 300

Jenkins, B. A. - 89

Jenney, J. - 189

Johnson, L. F. - 301, 302, 303, 304, 305, 306, 307, 308, 309, 310, 311

Johnson, R. E. - 312

Johnson, T. S. - 313

Jones, R. C. - 282

K

Kabota, K. - 314

Kahng, D. - 309

Kaiser, W. - 115, 214, 215, 216, 217, 315, 316, 317, 318, 319, 708

Kallman, H. - 320

Kamal, A. K. - 321, 322, 323

Kaminow, I. P. - 324, 325

Kanable, N. - 135

Kannelaud, J. - 133, 134

Kapany, N. S. - 400

Kaplan, J. I. - 327

Kaplan, R. A. - 326

Kassel, S. - 328

Kastler, A. - 329, 330

Katzman, M. - 331

Kaya, P. - 332

Keating, J. D. - 22

Keck, M. J. - 318

Keck, P. H. - 18, 333, 334, 335, 336, 337, 338

Kemeny, G. - 683, 684

Kessler, B. - 147, 148

Keyes, R. J. - 339, 340, 341, 530

Khanin, Ya. I. - 184

Kiang, Y. C. - 479

Kiel, A. - 342, 343

Kielch, S. - 515

Killpatrick, J. - 344

Kimura, T. - 562

Kindel, D. J., Jr. - 230

Kindlmann, P. J. - 38, 39

Kingsley, J. D. - 244, 345, 346

Kingston, R. H. - 347

Kisliuk, P. - 112, 113, 222, 348, 349, 350, 351, 363

Kiss, Z. J. - 176, 352, 353, 354, 355, 356, 357, 358, 359, 360, 444, 703

Klass, P. J. - 361

Klein, M. P. - 222

Kleinman, D. A. - 362, 363

Klemens, P. G. - 364

Kleppner, D. - 365

Knestrick, G. L. - 366

Koehler, T. R. - 367

Koester, C. J. - 368, 369, 370, 371

Kogelnik, H. - 73, 191, 372, 373, 547

Koozekanani, S. - 105, 374, 375, 376

Koster, G. E. - 377, 631, 632

Kotik, M. J. - 378

Krag, W. E. - 530

Kremen, J. - 379

Kroeger, R. - 412

Krokhin, O. N. - 26, 27

Kroll, N. M. - 380, 381

Krutchkoff, A. - 105, 375, 376

Kurokawa, K. - 562

L

Laff, R. A. - 87, 382

LaMarca, L. G. - 585

LaMarre, D. A. - 368

Landon, A. J. - 635, 636

Lankard, J. R. - 621, 627, 628, 634

Lasher, G. - 383, 384, 477

Lasser, M. E. - 270

LaTourrette, J. T. - 291

Laures, P. - 385

Lax, B. J. - 386, 387, 457, 530

Lax, M. - 254

Leifson, O. - 331

Leite, R. C. - 64, 388, 718

Lempicki, A. - 389, 564

Lengyel, B. A. - 82, 390, 391

Leontovich, A. M. - 206

SUBJECT INDEX

Outline of Classification System

A. Usage and Scientific Applications
- (1)   General
- (2)   Chemistry
- (3)   Communications and radar
- (4)   Geodesy and surveying
- (5)   Interferometry
- (6)   Materials processing
- (7)   Medicine and biology
- (8)   Optics and spectroscopy
- (9)   Physics and astronomy
- (10)   Weapons and military

B. Theory
- (1)   General electromagnetic radiation

Electromagnetic Radiation
- (2)   General
- (3)   Cavity, mode and interferometer
- (4)   Coherence

Interaction of Radiation and Matter
- (5)   General
- (6)   Absorption, emission and fluorescence
- (7)   Amplification and regeneration
- (8)   Energy levels
- (9)   Lifetimes or transition probabilities
- (10)   Modulation

Laser Action
- (11)   General
- (12)   Line width determinations
- (13)   Organic lasers
- (14)   Power (input and output) computations or analyses
- (15)   Pulsations (giants)
- (16)   Rate equation – threshold conditions
- (17)   Semiconductors and laser action

Nonlinear Effects of Light
- (18)   General

INDEX

B-8
257, 295, 296, 297, 342, 403, 553, 560, 700

B-9
342, 343, 377, 655, 658, 700

B-10
23, 48, 86, 98, 130, 131, 132, 143, 192, 233, 247, 248, 249, 321, 332, 412, 506, 551

B-11
22, 24, 25, 28, 52, 124, 128, 142, 144, 152, 156, 160, 177, 184, 235, 419, 501, 502, 503, 514, 518, 566, 568, 569, 570, 571, 575, 601, 602, 604, 643, 644, 648, 666, 667, 670, 671, 676, 716

B-12
169, 229, 239, 252, 326

B-13
78, 408, 564, 706

B-14
169, 193, 194, 200, 252, 323, 333, 334, 504, 558, 642, 650, 651, 680, 681, 682, 690, 691, 717

B-15
261, 391, 690, 691

B-16
26, 160, 204, 260, 327, 426, 483, 526, 538, 549, 587, 588, 589, 619, 669, 727, 731

B-17
46, 47, 171, 172, 173, 339, 383, 384, 457, 718

B-18
14, 170, 239, 283, 323, 516, 521, 550, 552

B-19
83, 347, 500, 529, 707

B-20
56, 169, 251, 262, 283, 567

B-21
208, 529

C-1
54, 127, 221, 255, 256, 273, 319, 328, 390, 392, 421, 599, 666, 729, 730

C-2
220, 456

C-3

11, 16, 30, 137, 152, 225, 315, 317, 349, 352, 353, 356, 396, 409, 415, 416, 417, 444, 465, 472, 603, 649, 661, 672, 673, 708, 712

C-4

63, 167, 187, 188, 189, 473, 492, 493, 528, 557, 578, 659

C-5

8, 12, 21, 33, 36, 37, 39, 41, 59, 65, 94, 120, 133, 134, 135, 185, 213, 234, 235, 240, 241, 250, 256, 263, 266, 267, 268, 269, 272, 288, 298, 299, 300, 330, 344, 365, 372, 385, 405, 407, 445, 446, 447, 508, 510, 511, 512, 527, 531, 532, 544, 584, 600, 607, 693, 694, 695, 726, 731

C-6

31, 51, 67, 71, 79, 122, 153, 215, 216, 217, 222, 301, 302, 303, 304, 305, 306, 307, 308, 310, 311, 314, 316, 317, 336, 337, 356, 357, 358, 359, 360, 386, 415, 416, 417, 418, 420, 422, 423, 424, 462, 463, 464, 485, 522, 523, 524, 525, 572, 573, 574, 622, 623, 624, 625, 626, 627, 628, 697, 698, 709, 719

C-7

209, 210, 211, 212, 389, 437, 438, 469, 470, 612, 614, 615, 616, 618, 701, 710, 721, 722

C-8

27, 40, 53, 64, 87, 88, 89, 90, 92, 163, 174, 186, 244, 274, 280, 281, 284, 285, 340, 341, 345, 346, 382, 387, 406, 459, 460, 467, 474, 475, 476, 477, 485, 530, 621, 634, 661, 662

C-9

43, 154, 561, 720

C-10

35, 38, 50, 58, 61, 68, 69, 70, 72, 73, 76, 85, 104, 106, 116, 117, 118, 119, 121, 136, 191, 203, 207, 216, 223, 228, 236, 242, 243, 256, 267, 268, 275, 282, 287, 312, 331, 350, 363, 367, 368, 373, 378, 379, 393, 399, 405, 434, 436, 442, 443, 448, 468, 495, 519, 537, 545, 546, 547, 555, 565, 585, 586, 593, 607, 613, 617, 629, 646, 664, 675, 683, 684, 689, 711, 726, 728

C-11

18, 81, 100, 101, 102, 103, 108, 109, 190, 218, 333, 334, 335, 336, 337, 338, 360, 461, 485, 487, 544, 565, 592, 598, 635, 636, 652, 675, 722

C-12

138, 276, 440, 441, 678

C-13

114, 115, 202, 272, 318, 319, 351, 620

C-14

37, 71, 123, 236, 280, 306, 307, 308, 336, 375, 397, 480, 715, 721

C-15

60, 246, 249, 517, 594, 595

C-16

9, 33, 62, 75, 86, 125, 126, 205, 221, 241, 270, 290, 291, 374, 376, 449, 450, 454, 455, 562, 563, 696

C-17

3, 9, 19, 95, 139, 178, 189, 227, 234, 318, 336, 337, 340, 348, 374, 376, 388, 394, 559, 576, 610, 694

C-18

15, 29, 163, 201, 224, 251, 292, 293, 294, 300, 316, 398, 423, 424, 462, 463, 464, 608, 609, 631, 632, 656, 657

C-19

1, 2, 4, 5, 7, 17, 20, 44, 45, 55, 66, 74, 105, 107, 110, 111, 129, 145, 146, 147, 148, 157, 158, 159, 176, 180, 181, 182, 191, 206, 214, 216, 242, 256, 264, 265, 266, 267, 268, 270, 281, 317, 326, 341, 345, 346, 351, 382, 393, 399, 400, 413, 414, 427, 435, 459, 478, 482, 486, 489, 498, 513, 534, 535, 542, 543, 591, 596, 597, 630, 633, 634, 638, 639, 640, 641, 645, 647, 653, 665, 677, 679, 685, 688, 713, 714, 725

C-20

32, 61, 84, 86, 112, 113, 149, 151, 205, 243, 245, 246, 249, 284, 285, 290, 309, 324, 325, 354, 355, 371, 431, 451, 452, 479, 488, 496, 497, 499, 594, 652

C-21

133, 134, 225, 481, 498, 631, 632

C-22

65, 94, 99, 150, 151, 179, 238, 258, 428, 432

C-23

164, 165, 240, 621, 703

C-24

6, 7, 12, 91, 157, 158, 159, 303, 322, 341, 355, 364, 652, 702, 720

C-25

34, 281, 312, 340, 341, 342, 428, 433, 507, 699

C-26

226, 712